ROOF CONSTRUCTION
AND LOFT CONVERSION

THIRD EDITION

C. N. Mindham

Blackwell
Science

© C.N. Mindham 1986, 1988, 1994, 1999

Blackwell Science Ltd, a Blackwell Publishing Company
Editorial Offices:
Osney Mead, Oxford OX2 0EL, UK
  Tel: +44 (0)1865 206206
Blackwell Science, Inc., 350 Main Street,
Malden, MA 02148-5018, USA
  Tel: +1 781 388 8250
Iowa State Press, a Blackwell Publishing Company,
2121 State Avenue, Ames, Iowa
50014-8300, USA
  Tel: +1 515 292 0140
Blackwell Science Asia Pty, 54 University Street,
Carlton, Victoria 3053, Australia
  Tel: +61 (0)3 9347 0300
Blackwell Wissenschafts Verlag,
Kurfürstendamm 57, 10707 Berlin, Germany
  Tel: +49 (0)30 32 79 060

First edition published under the title *Roof
  Construction for Dwellings* by Collins Professional
  and Technical Books 1986
First Edition revised published by BSP Professional
  Books 1988
Reprinted 1989
Second Edition published by Blackwell Science 1994
Reprinted 1995, 1996 (twice), 1997, 1998
Third Edition published 1999
Reprinted 2000 (twice), 2001, 2002

Library of Congress
Cataloging-in-Publication Data
Mindham, C.N. (Chris N.)
    Roof construction and loft conversion/
C.N. Mindham.—3rd ed.
      p. cm.
    Includes bibliographical references and index.
    ISBN 0-632-05201-5 (pb)
    1. Roofs—Design and construction.   2. Framing
(Building).   3. Lofts—Remodeling.   I. Title.
TH2393.M63   1999
  695–dc21   99-13158
                                                        CIP

ISBN 0-632-05201-5

A catalogue record for this title is available
from the British Library

Set in 11/13pt Plantin
by DP Photosetting, Aylesbury, Bucks
Printed and bound in Great Britain by
MPG Books Ltd, Bodmin, Cornwall

For further information on
Blackwell Science, visit our website:
www.blackwell-science.com

Contents

Preface

When my book on the construction of roofs for dwellings was published in 1986, it filled a real gap in the literature as the first reference manual on the subject. Discussion with readers and further research revealed a demand for more information on the conversion of loft space to usable attic accommodation, and a second edition was published in 1988 to include guidance on the subject.

The third edition takes account of the design and construction changes required by British Standard 5268: Part 2: 1996 and BS 5268: Part 3: 1998, both issued since the book was last revised. It also introduces the reader to the forthcoming Eurocode requirements.

Like many other building topics the roof is one of those subjects with which everyone is familiar until it comes to actually detailing or cutting the timber components concerned, and then the lack of knowledge becomes apparent. Furthermore, research soon confirmed the total lack of in-depth text on the construction of trussed rafter roofs, a method of construction now used on over 90% of house construction in the United Kingdom.

The book aims to describe with the aid of many drawings, not the structural design analysis of the roof structure, but the design of the roof assembly as a whole entity rather than individual elements in isolation. Recognising the growing trend to refurbish older homes, the traditional or 'cut' roof is described. The bolted and connectored roof is dealt with in some detail, for despite the popularity of the trussed rafter this older system is still chosen by some builders. The bolt and connector truss roof is particularly popular for small extension projects where it often continues the construction of the original roof.

Chapters 5 and 6 cover the trussed rafter roof in great detail, dealing with the often misunderstood hip construction, valleys, girder truss assemblies, and the forming of openings in roofs as well as attic constructions. Chapter 6 compares the various truss plate systems and has been made as accurate as possible, bearing in mind the many changes being introduced by these manufacturers to their engineering services and computer programs and with the constant updating of BS 5268: Parts 2 and 3 and Eurocodes.

Chapters 8 to 11 deal with all aspects of loft conversion to attic rooms of the roof structure itself. The text does not address the subject of fire protection and escape, or the installation and alteration to services. Variations between buildings being converted in shape of roof, size, number of storeys, and intended use of attic are so great that it is impossible to cover all situations likely to be encountered. My text and illustrations will, however, cover most common constructions.

It is the intention that the book be used for reference, and to this end there is a small degree of repetition between chapters, and there is frequent cross-referencing between chapters for both text and illustrations. Although some basic common knowledge of building is anticipated, most terms used are fully described, making the book equally suitable for use by both the building student and the professional. The text takes into account the latest issues of both the British Standard for timber engineering, BS 5268: Parts 2 and 3, and the Building Regulations 1991 and all subsequent amendments. However, as it was felt to be outside the scope of this book, the subject of fire resistance and spread of flame has not been dealt with. Reference should be made to Building Regulation Approved Documents.

For ease of reference all drawings have been given a number, the first digit of which refers to the chapter, and the second and third digits being the numerical sequence in that chapter. Generally, shading has been used to highlight those elements discussed in the text to which the illustration applies. Most drawings have been produced in perspective form to aid quick appreciation of the three-dimensional nature of all roof structures. Chapter 2 sets out the terms used throughout the book to describe roof and truss shapes, and individual roof members. The specialised terminology of the trussed rafter is given in Fig. 5.2. Finally for those involved in the design aspects of roof structures, the British Standard 5268: Parts 2 and 3 should be available for ready reference.

C.N. Mindham
14 Harrowden Lane
Finedon
Northants
NN9 5NW

ACKNOWLEDGEMENTS

I would like to thank Martin Moore of Wolf Systems Limited, who has been of great help with the structural design aspects of this revised edition, and all who, however fleetingly, have helped me with the research and preparation of this book. I would also like to thank some of the purchasers of the first two editions, who have taken the trouble to telephone me and discuss various aspects of both text and illustrations: some of the changes I have made in this third edition have been in response to their suggestions.

Chapter 6 has been produced with the willing co-operation of Messrs. Wolf Systems Limited, Gang-Nail Systems Limited, MiTek Industries Limited, and Alpine – Twinaplate Limited, all of who have given freely of their technical information and to whom I am indebted for the use of their various illustrations in this chapter. Finally my thanks are due to my wife for tolerating the sometimes not inconsiderable mess of paper, literature and drawings cluttering the family home.

CHAPTER 1

The Development of the Pitched Roof

PRIMITIVE ROOF FORMS

Man has always needed a roof for shelter. Early man used roofs formed by nature such as caves, but nomadic peoples had to be more resourceful, creating shelters of a temporary nature each time they moved. It is likely that simple tents formed with animal skins over branches were the early form of constructed roofs, with more permanent shelter being pit dwellings. These were simply a shallow excavation covered with a simple roof of branches and skins. It is an easy step from this type of dwelling to a simple wall on the edge of the pit to raise the headroom and then to use shaped branches to give a slight pitch, thus improving rain run-off and therefore the quality of the environment within the shelter.

The simple 'cruck' frame comprised two curved pieces of timber standing on the ground at one end and meeting at the top. Across several of these 'crucks' were tied horizontal members onto which, again, were fixed skins or as time progressed simple thatch.

THE COUPLED ROOF

Moving away from early roof forms that provided both wall and roof in one unit, the next development showed a true roof built on masonry or timber walls. The simplest form of roof was a coupled roof, consisting of two lengths of timber bearing against each other at the top and resting on a wall plate at their feet. The timbers, called couples, were pegged together at the top with timber dowels and were similarly pegged or spiked to the wall plate. The term 'couple' was used until the fifteenth century when the terms 'spar' or 'rafter' started to be used. The term rafter of course is still used to describe the piece of timber in a roof spanning from the ridge to the wall plate.

The couples were generally spaced about 400 mm apart tied only by horizontal binders and tile battens. The simple couple was adequate for small span dwellings and

1

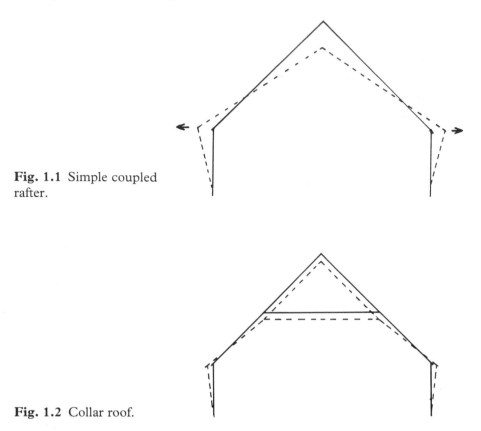

Fig. 1.1 Simple coupled rafter.

Fig. 1.2 Collar roof.

steep pitches, but the outward thrusting force at the feet of the rafters caused stability problems with the walls, and excessively long rafters sagged in the middle under the weight of the roof covering. The illustration in Fig. 1.1 shows the required shape in solid line and the deflected shape in dotted line.

To overcome both of these problems the 'wind beam' or 'collar' was introduced. Whether the collar acts as a tie or a strut for the couples will depend upon the stiffness of the supporting wall below. Assuming however, that the wall is so substantial that it will not be pushed outward by the bottom section of the couples, then the collar will act as a strut. If however, as is more likely with early timber framed buildings, the wall is relatively flexible then in that case the collar would act as a tie holding the couples together. There would still be some outward thrust but this would be limited by the collar to the degree of bending in the lower part of the couple only. Figure 1.2 illustrates this condition. It can readily be appreciated that in larger roofs, where the walls are relatively flexible, there is a considerable tying effect in the collar demanding a more sophisticated joint between collar and couple than could be achieved with simple iron nails. The collar was therefore frequently jointed to the couple with a halved dovetail shaped joint, often secured with hardwood pegs.

STABILITY

The next development was to fit additional members to assist with the stability of the roof in windy conditions and these were called 'sous-laces' or braces. On roofs constructed on substantial masonry walls which were also very thick, further struts or 'ashlars' were introduced to stiffen the lower section of the couple. Figure 1.3 illustrates this form of construction, the wall plate being well fixed to the wall with the bottom member of the ashlar halved over it to prevent the roof sliding on the top of the wall.

These now very substantial 'couples' began to be spaced further apart and became known as 'principals'. Between these main members simple couples or 'rafters' were placed, but to avoid sag or to accommodate longer rafter length possibly not available in one length of timber, an intermediate support was needed and this was called a 'purlin'. The purlin is in turn supported by the principal couples, as shown in Fig. 1.4.

The tendency for the roof to spread was now concentrated in the heavily loaded principals and it became apparent that if spans were to increase this spreading would have to be controlled. The 'tie beam' was introduced thus forming the first 'trussed' or 'tied' roof. Figure 1.5 illustrates the roof form described.

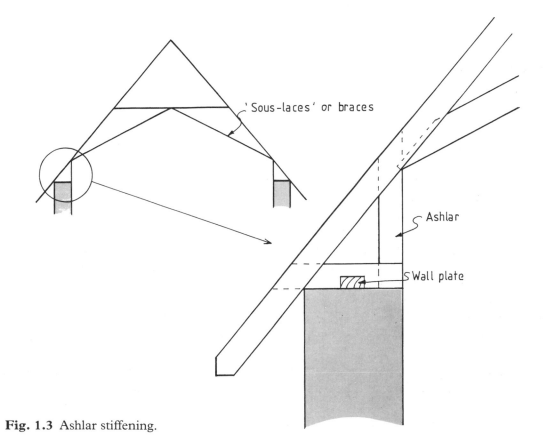

Fig. 1.3 Ashlar stiffening.

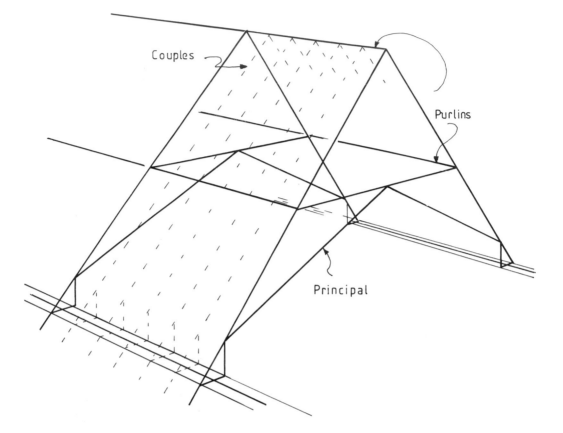

Fig. 1.4 Principal truss and purlin roof.

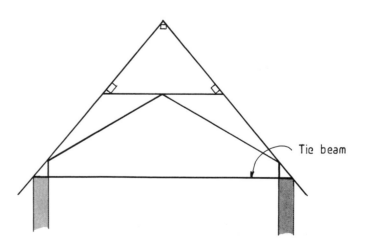

Fig. 1.5 Tie beam truss.

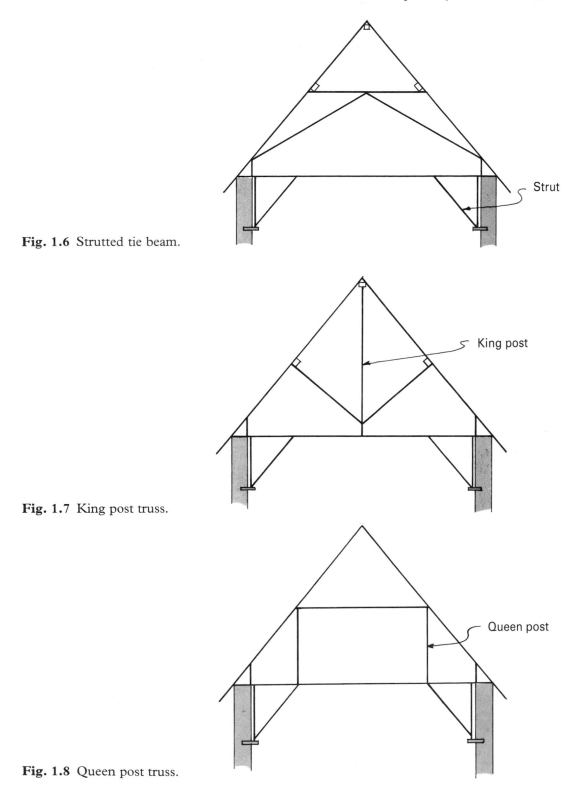

Fig. 1.6 Strutted tie beam.

Strut

Fig. 1.7 King post truss.

King post

Fig. 1.8 Queen post truss.

Queen post

As development progressed the span of the roof was limited only to the availability of long timbers used for the tie beam, but it is obvious that these long beams themselves would tend to sag under their own weight. To prevent this happening they too had to be supported and this was done with the introduction of 'struts' fitted to a corbel built into the wall below, as illustrated in Fig. 1.6.

With this tie beam now becoming a major structural member a different configuration of members evolved becoming more like the truss common today. Having stiffened the tie beam it became apparent that this could be used as a major structural item from which to support the principals. The major support running from the centre of the tie beam to the ridge purlin was known as the 'mountant' now referred to as a 'king post' (see Fig. 1.7). A king post truss is also illustrated in Fig. 8.1, being used as part of the structure of an attic room. With two posts introduced the roof form is known as a 'queen post' truss, which in its simplest form is shown in Fig. 1.8. This particular roof form gave the opportunity of providing a limited living space within the roof. It should be remembered that until this stage of development all roof forms and trusses described had no ceiling and were open to the underside of the rafters and roof covering. To use the queen post roof form as an attic, a floor was needed thus creating a ceiling for the room below.

CEILINGS

Ceilings were first referred to in descriptions of roofs in the fifteenth century when they were known as 'bastardroofes' or 'false roofs' and then later as 'ceiled roofs', hence 'ceiling' as we know it today.

The ceiling supports were known as joists or cross beams again being supported by the hard working tie beam between the principals. The construction is illustrated in Fig. 8.2.

Continuing developments of the roof form itself, and demand for even larger spans and heavier load resulted in some relatively complex principals or trusses being developed. One such form was the 'hammer beam' roof, illustrated in Fig. 1.9. Clearly this is not a roof to be 'ceiled', being very ornate as well as functional.

The hammer beam roof is generally to be found supporting the roof over halls in large mansions and of course churches. The roof was framed in such a way as to reduce the lateral thrust without the need for a large and visually obstructing tie beam. The walls onto which such a roof was placed had to be substantial and were often provided with buttresses in line with the principals to contain any lateral thrust that may develop.

TRUSSES

Roofs in truss form developed using carpentry joints and some steel strapping, until the latter part of the eighteenth century when bolts, and even glues, started to be used to create large truss forms from lighter timber members. Such truss forms often used

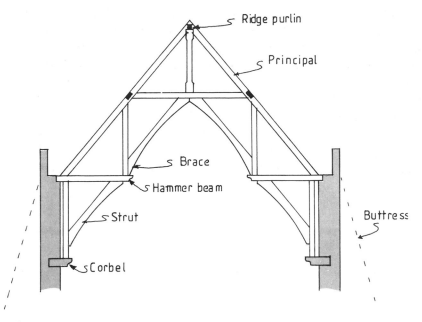

Fig. 1.9 Hammer beam truss.

softwoods, as distinct from the hardwoods more frequently used in the shapes previously described. The large timber sections in oak particularly were becoming very scarce and of course very expensive. Whilst some significant advances in span were achieved, using the techniques described above, the domestic roof did not require very large spans and changed very little from the collared coupled roof. Indeed many small terraced houses built during the eighteenth and nineteenth century required no principals at all. The dividing walls between the houses were close enough to allow the purlins to rest on these walls, effectively using them as principals. Figure 1.10 illustrates a typical terraced house roof construction.

The larger properties where the span of the purlin was too long for one piece of timber, or where hip ends were involved, continued either to use the established methods of construction using principals, collars and purlins, but it was common practice to omit the principals and to support the purlins off the walls below with posts or struts.

DESIGN FOR ECONOMY

In 1934 the Timber Development Association (TDA) was formed, now known as TRADA (Timber Research and Development Association). The Association took up the work already being done at that time by the Royal Aircraft Establishment and progressed work on timber technology alongside the Forest Product Research Laboratories. Although the Royal Aircraft Establishment may sound a strange body to be interested in timber, it must be remembered that many aircraft of that era, and some notable ones after such as the Mosquito, used highly stressed timber structures for the

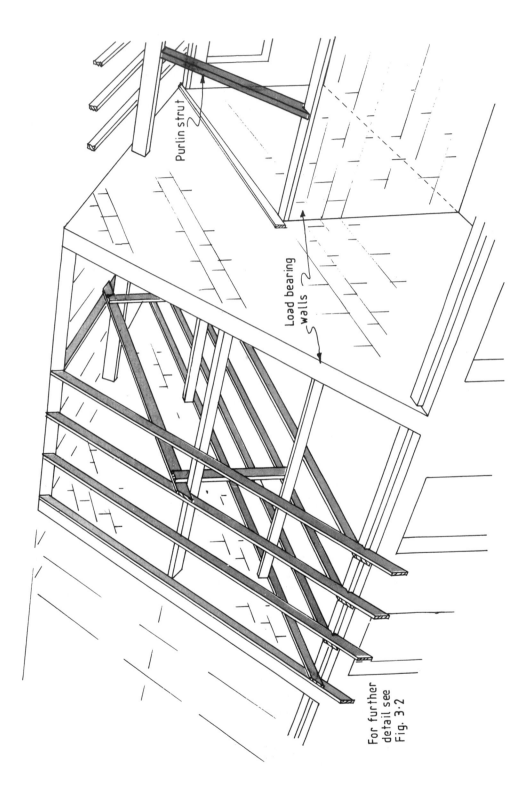

Fig. 1.10 Purlin and common roof.

fuselage and wings. Some aircraft hangars were of timber construction and utilised record breaking large span small timber section trusses with bolted joints.

After the Second World War shortages of materials resulted in a licence being required for all new building works, making economy in use of paramount importance. Imported materials such as timber were very much at a premium and TDA was given the task to find ways of economising on the country's use of timber. Quite correctly they identified the roof structures of buildings as a high volume user of timber and developed a design for a domestic roof using principal trusses constructed of small timber sections connected with bolts and metal connector plates. The roof used purlins and common rafters similar to the systems previously discussed. These trusses became known as 'TDA' trusses, and with some minor modifications are still in use today. It appears that some of these designs were available shortly after the Second World War but were first published as a set of standard design sheets around 1950.

The designs were based on existing truss shapes but were not engineered in the sense that structural calculations were prepared for each design. Load testing on full size examples of the truss was used to prove their adequacy and from these tests other designs developed.

STANDARD DESIGN ROOFS

The first designs produced were known as 'A' and 'B' types, dealing with 40° and 35° pitches respectively. They covered spans up to 30 ft (9 m).

House design fashion changed during the later 1950s and early 1960s, demanding lower roof pitches. 1960 saw the introduction of the TDA type 'C' range for pitches between 22° and 30°. Spans were also increased up to 32 ft (10.8 m). Around 1965 the types 'D', 'E' and 'F' ranges were published; these later designs using a slightly different truss member layout went down to 15° pitch and up to 40 ft (12 m) span. Further designs used trusses spaced at 6 ft (1.8 m) centres and had some degree of pitch and span flexibility within specified limitations.

A range of designs for trussed rafters (i.e. each couple tied together at ceiling level) was produced also using bolt and connector joints, but these were designed only to carry felt roof coverings and did not prove as popular as the principal truss designs. Industrial roofs were not neglected, with principal truss designs using the bolt and connector joint techniques for pitches of 22.5° spacing between 11 and 14 ft (3.35–4.25 m) and up to 66 ft (20.1 m) span.

Whilst roofs are still constructed using these techniques, the TDA designs are no longer available from TRADA.

BOLT AND CONNECTOR JOINTS

All of the TDA principal and trussed rafter designs used bolts and connectors at joints where previously mortice and tenon, half lap or straight nailed or pegged joints would

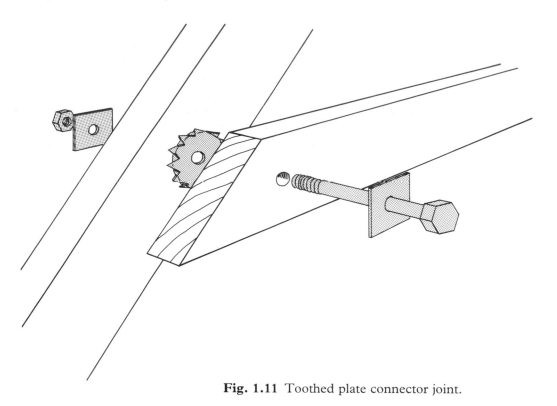

Fig. 1.11 Toothed plate connector joint.

have been used. The small timber sections used in the designs of the trusses did not allow the use of conventional carpentry joints and gave insufficient nailing area for an all nailed assembly. The connector allows the forces in the joint to be spread over a large area of the connected timber, the bolt holding the timbers in place thus allowing the connector to transmit the load from one truss member to the other. Figure 1.11 illustrates the typical single connector joint.

TRUSSED RAFTERS

In the early 1960s the punched metal connector plate was introduced into the UK from the USA and was to revolutionise the construction of domestic roofs even more than the TDA truss designs described. There are now four main plate manufacturers in the UK, the first in 1967 being Gang-Nail whose name has come to be used to describe all punched metal connector trusses, in the same way that 'Hoover' seems to describe a vacuum cleaner.

Trussed rafters are generally prefabricated in a factory and transported to site, although with certain types of plate, fabrication can take place on site. In the case of metal plates, the manufacturer sells plates backed up to varying degrees with design aids to approved manufacturers, many of whom are also timber merchants. The timber used is both graded for strength and machined on all surfaces to give accuracy to the finished product. Trussed rafters can also be assembled using plywood gussets, the plywood being either nailed to a defined pattern or nailed and glued to the truss members to form the joint. Ply gusseted trusses are not as popular as metal plated trusses, but do offer a method of manufacture not requiring specialist equipment. Similarly the galvanised steel plates punched with a pattern of holes to receive nails can also be used to form truss joints and these too can be fabricated on site.

The punched metal nail plates used in factory trussed rafter production are mechanically pressed into the timbers on both sides of each joint to form a trussed rafter. This trussed rafter is then placed on the roof at approximately 600 mm centres taking the place of the common rafter. Hence its term 'trussed rafter', as distinct from the TRADA type principal truss, although it will be seen later in Chapter 5 that trussed rafters themselves can be used to form principal or girder trusses. A typical 'fink' trussed rafter is illustrated in Fig. 1.12.

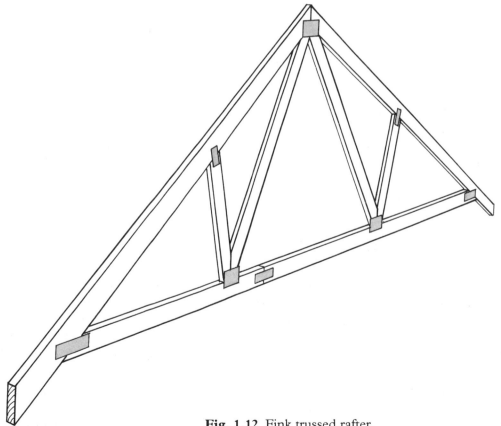

Fig. 1.12 Fink trussed rafter.

COST ADVANTAGES

Trussed rafters are designed to carry simply the direct load imposed upon them. It is assumed that they are to be kept upright by other members, these members being the binders and diagonal bracing and even the tile batten vital to the overall stability of the roof. Whilst most trussed rafters are used for roofs of housing, their use is increasing for roofs of public buildings, commercial buildings and to a lesser extent for industrial and agricultural buildings. Clear spans in excess of 30 m can be achieved with lightweight roof coverings.

When first introduced into the UK, the designs were limited to those contained in standard design manuals, thus the duo pitch and mono pitched roofs were common but more complex roofs needed individual designs prepared. The advent of the computer both speeded up and dramatically reduced the cost of the design process, and this has been further advanced by the use of microcomputers installed in all trussed rafter manufacturers' offices. There are now almost no limitations to the possible shape of trussed rafters, except those imposed by the practicality of production and transportation to site. The power of computers enables not only the individual trussed rafter to be designed but also the whole roof as a structural entity. Roof layout drawings can be produced in minutes and then either plotted on to paper or sent via 'e-mail' to the end user.

Trussed rafter roofs use approximately 30% less timber than a traditional roof, and can be built into a roof form in a fraction of the time taken for either a truly traditional common and purlin roof, or a TRADA construction. Factory production keeps the labour cost of trussed rafter manufacture very low compared to that necessary to assemble a bolt and connected jointed truss, thus giving further cost advantages to the trussed rafter. Almost all new housing now uses a trussed rafter form of roof construction.

LEGISLATION

Having looked at the development of the roof form, we must take account of the legislation controlling building construction in the UK. Before this century no controls existed, and it was not until the introduction of the model byelaws by each local authority area that some degree of control was placed upon the design of buildings.

The Building Regulations as we now know them first appeared in 1965, and have been amended and re-issued on several occasions since that date. Subsequent amendments have dealt with such roof related matters as the restraint of gable and walls, thermal insulation and roof void ventilation. The first major change to the Building Regulations occurred in 1985 and took the form of a two-part publication, the first part setting the standards to be achieved and the second, for approved documents, laying down approved methods of achieving them. The third edition of this book has been produced in the light of the latest edition of the Building Regulations which came into force in 1991 including the recent amendments. These regulations lay down the legal requirements for building and concern themselves with health and safety aspects and not the aesthetic aspects of the structure. The latter, of course, is controlled by the local planning authorities.

The National House-Building Council (known as NHBC) has its own set of standards, which although incorporating the Building Regulations requirements, look beyond the health and safety aspects and seek to set minimum standards for quality control and such items as heating, electrical power sockets, and the general finish given to the buildings. Formed in 1936 it was not until the mid-1960s that the council began to have influence on the vast majority of house builders in the UK.

Concerned by the so-called 'jerry builders' after the Second World War, the building societies needed some method of ensuring that the homes on which they had granted mortgages were of an adequate standard to protect their investment. These societies therefore demanded that house builders building and wishing to sell homes on which the societies were granting mortgages must belong to the NHBC and submit themselves to their inspections. Having achieved full compliance with the NHBC requirements and of course the Building Regulations, the mortgage would be granted. Consequently most newly built homes until now have had to be inspected by the local authority as well as the NHBC, although this is likely to change in the near future, and only the inspectorate of the NHBC will be involved. An alternative to NHBC for mortgage purposes in most instances, is that the house should be inspected by a registered architect, and this seems to be the only way that a non-registered house builder can build and sell a new home under a mortgage agreement.

The Building Regulations and NHBC standards in turn refer to various *British Standards* and it is intended here only to deal with those British Standards concerned with timber in roof structures.

Code of Practice 112 started life in 1952, and was amended in 1967 when the principle of allocating grade stresses to timber was introduced. 1971 saw further changes to the code of practice, then issued with stresses and timber sizes in metric units. This code became British Standard 5268 which itself was split into many parts: Part 2 deals with the general principles of timber structural design, whilst the latest edition of Part 2: 1996 has simplified the hitherto relatively complex subject of stress grading by grouping timbers into strength classes ranging from C16 for softwoods to D70 for hardwoods. However, the Building Regulations approved document table still refers to the earlier issue of BS 5268: Part 2: 1991, and remains based on SC3 and SC4 grades.

The current standard recognises a special grade for the use in punched metal nail plated trussed rafters known as TR6. BS 5268: Part 3: 1998 deals specifically with the design and fabrication of trussed rafter roof construction. BS 6399: Part 3: 1997 is the code of practice for the loads imposed on roofs, dealing with such aspects as dead and live loads as well as snow loading. BS 5250 concerns itself with the roof void ventilation and was last reviewed in 1995. BS 5534: Part 1: 1997 deals with the design of slating and tiling with the recommendations for workmanship for these roof coverings being given in BS 8000: Part 6.

British Standard 4471: 1987, *Sizes for sawn and processed softwood* has now been withdrawn and replaced by an English language version of the European code EN 313, known as BSEN 1313/1: Part 1: 1997, *Softwood sawn timber*. The standard sets out standard sizes and processing tolerances, whilst BS 4978, revised in 1996, deals with the stresses allocated to structural timbers. This edition has been revised to take account of the publication of the relevant European Standards:

(1) Changes have been made to the visual grading section in accordance with BSEN 518 structural timber – grading – requirements for visual standards.
(2) Machine strength grading is now specified in BSEN 519 structural timber – grading – requirements for machine strength graded timber and grading machines. The sections concerning machine strength grading have been deleted and the title has been changed.
(3) The sections concerning visual strength grades for laminations have been deleted.

This British Standard specifies the means of assessing the quality of softwoods for which grade stresses are given in BS 5268: Part 2. This document is recommended for those wishing to have some insight into the visual appearance of the type of timber that they can expect with the various stress gradings. Such factors as knots, fissures, bow, spring and twist are dealt with, giving limiting factors.

The above deals with timber from European countries. Timber from Canada and the USA is covered by their own standards which are recognised in the UK for visually graded timber. These are NLGA, Canada, 1994 national grading rules for dimension lumber, and NGRDL, USA, 1997 national grading rules for softwood dimension lumber. There is also a machine grade standard known as NAMSR 1986 set by the North American exports standard. This was introduced to give more precise selection of strength potential, thus increasing the economic use of this natural resource.

All structural timber used in dwellings must now be graded into stress limiting classes and marked with the grades. The mark must show not only the grade, but the grader and the grading station, the British Standard number and the species group. Alternatively of course it can be marked with the approved Canadian and American grading stamps. Grading can be carried out either visually by qualified visual graders, or by licensed stress grading machines operated by trained staff.

Earlier standards classified timbers within a single species and were developed from an assessment of the timber's strength compared to that of a defect free sample, thus the old grades of 40, 50, 65 and 75 represented the percentage strength of the sample compared to the defect free sample. Thus with different species offering different strength properties it can be seen that a weak timber (say Balsa wood), at 75 grade would be much weaker than British pine at 75 grade. The strength classes simplify this by classifying by strength regardless of species, thus a piece of C14 balsa (not that it actually exists), would have the same structural ability as a piece of C14 British pine. This of course simplifies design unless visual appearance is of importance on exposed structural feature members of a roof form, in which case the designer should refer to BS 4978 to gauge for himself the visual defects likely to be allowable under the strength class selected by structural analysis.

As old strength classes are still allowed by the current Building Regulations 1991, the comparison table below may be of interest and assistance.

- SC3 now C16;
- SC4 now C24;
- SC5 now C27.
- Softwood grades now range as follows: C14, C16, C18, C22, C24, C27 (and TR6 for trussed rafters), C35 and C40.
- Hardwood grades are now: D30, D35, D40, D50, D60 and D70. Most roof structural softwood timber will fall in the range C16/C24 plus of course TR26 for trussed rafters.

The second edition of this book at this juncture mentioned the new European Standards. It is true to say that some have been introduced, such as Euro code 5 and harmonisation has continued between British and European codes, hence the BSEN prefix now being used. However, at the house extension, alteration and repair end of the market, the SC3–SC5 strength classes still apply and are commonly in use.

As can be seen from the short discussion above, timber grading is a detailed scientific subject well outside the scope of this book. The book is concerned with the constructional design of the roof, rather than the calculation of structural design and specification. The reader is directed to those British Standards referred to above, the many publications available from TRADA, and the *Timber Designers' Manual* by Baird and Ozelton.

Roof development will undoubtedly continue. The timber sizes used in modern trussed roof construction really constitute the practical minimum possible. Structurally it may be feasible to reduce those sizes, but for reasons of achieving adequate fixings for ceiling boards and tile battens, the timber sizes cannot be reduced. It is therefore difficult to see beyond the trussed rafter, but its method of construction into the completed roof form may change.

Although the labour involved in erecting a trussed rafter roof is relatively small, access at roof plate level and within the roof structure whilst under construction is not good. As will be seen in Chapter 5, the British Standard requires a considerable amount of additional bracing to be installed within the roof, thus increasing the labour involved and the hazards of gaining access within the roof void. For this reason a practice relatively common with the large panel timber framed housing systems may become increasingly popular. This is to construct the roof including the wall plate at ground level, complete with all binders, bracing, ties, tank platform, tank, felt battens, barge and fascias where appropriate. This whole, relatively light assembly can then be craned on to the shell and fixed in position. It is not suggested that this is a cost effective method for very small building sites, but on the larger estates, where continuity of house building is achieved, it has many advantages, not the least of which is the safety of the workman concerned.

CHAPTER 2
Roof Shapes and Terminology

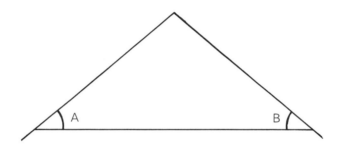

Fig. 2.1 Duo pitched roof. This is the most common roof shape with equal pitches on either side, i.e. angle A equals angle B.

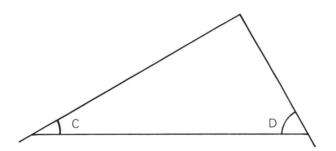

Fig. 2.2 Asymmetric roof; angle C is not equal to angle D.

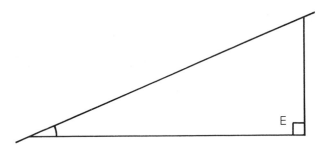

Fig. 2.3 Mono pitched roof; angle E equals 90°.

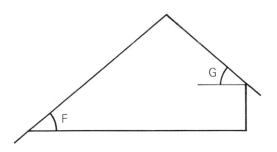

Fig. 2.4 Truncated duo pitched roof; angle F equals angle G. This truss form is often introduced into domestic housing in conjunction with the conventional duo pitched roof to form an interesting roof line.

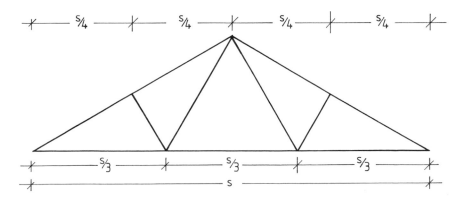

Fig. 2.5 Fink truss shape. This is the most common trussed rafter form used on spans of up to 8 to 9 m.

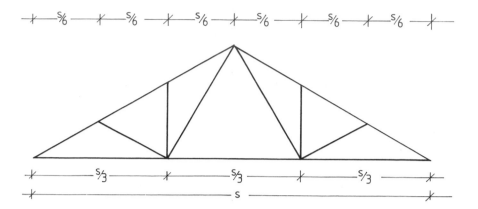

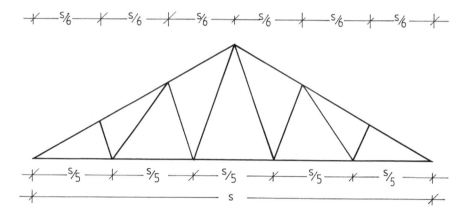

Fig. 2.6 Fan truss shape. This is used on larger spans and is a common trussed rafter form.

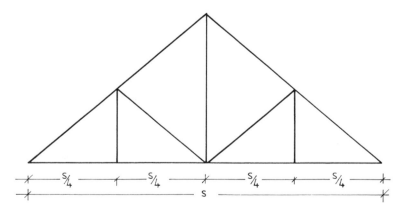

Fig. 2.7 Double 'W' shape. This is used on spans above 14 m and is not often used on housing.

Fig 2.8 Howe four bay truss. This is often used in trussed rafters in girder form. This could also be used in six bay configuration.

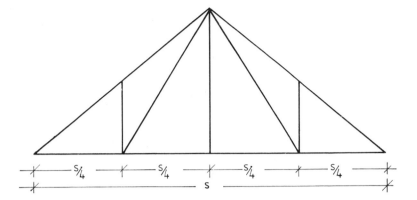

Fig. 2.9 Pratt four bay truss. This is occasionally used in trussed rafters in girder form.

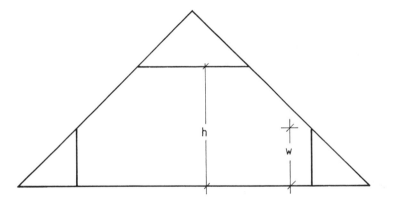

Fig. 2.10 Attic or 'room-in-roof' truss shape. This is a popular shape in trussed rafters: there are no minimum heights set for h and w, but for h a 2.3 m minimum is recommended, with 1.2–1.5 m being the practical minimum height for w.

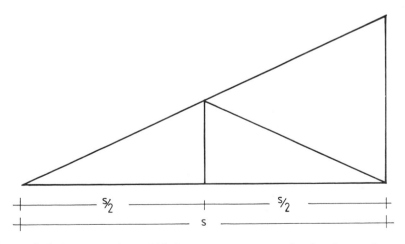

Fig. 2.11 Mono pitch truss two bay. This is a common trussed rafter form often used in conjunction with trusses in Figs 2.5 to 2.7.

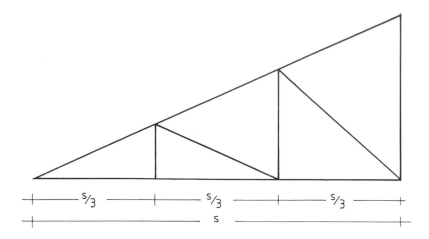

Fig. 2.12 Mono pitch truss three bay. This is similar to the mono pitch truss two bay (Fig. 2.11), but is suitable for larger spans.

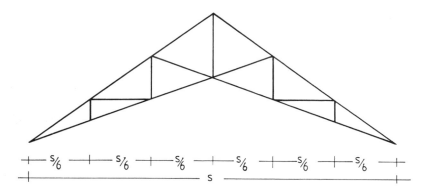

Fig. 2.13 Scissor truss. This is a possible trussed rafter shape occasionally used to create a feature ceiling in the lounge of a house.

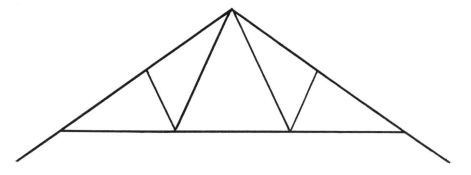

Fig. 2.14 Raised tie truss also used to create feature ceilings. Often fink based with rafters extended down to the wall plate.

TERMINOLOGY

Chapter 1 has given the history and derivation of some of the names given to roof structure members. The list below, although not exhaustive, describes the terms used on domestic roof structures.

The reader is referred to Fig. 2.15.

A – Wall plate – sawn timber, usually 50×100 or 50×75 mm bedded in mortar on top of the inner skin of a cavity wall. Straps must be used to secure the wall plate to the structure below (see Chapter 7, Figs 7.8 and 7.9).

B – Common rafter – sawn timber placed from wall plate to ridge to carry the loads from tiles, snow and wind. Long rafters may need intermediate supports from purlins.

B1 – Jack rafters – sawn timber rafter cut between either a hip or valley rafter (see Chapter 3, Fig. 3.7).

C – Ceiling joist – sawn timber connecting the feet of the common rafter at plate level. The ceiling joist can also be slightly raised above the level of the wall plate, but this would technically then be termed a collar. The ceiling joist supports the weight of the ceiling finish (normally plasterboard) and insulation. It may in addition have to carry loft walkways and water storage tanks, in which case it must be specifically designed to do so.

D – Ridge – a term used to describe the uppermost part of the roof. The term is also used to describe the sawn timber member which connects the upper parts of the common rafters.

E – Fascia – usually a planed timber member used to close off the ends of the rafters, to support the soffit M, to support the last row of tiles at the eaves N and to carry the rainwater gutter support brackets.

F – Hip end – whereas a gable end O is a vertical closing of the roof, the hip is inclined at an angle usually to match the main roof.

F1 – Hip rafter – sawn timber member at the external intersection of the roof slope (similar to a roof sloping ridge), used to support the jack rafters forming the hip (see Chapter 3, Fig. 3.7).

G – Valley – term used to describe the intersection of two roofs creating a 'valley' on either side. The illustration has only one main valley, the building being L-shaped on plan. A further small valley is illustrated on the dormer roof with its junction to the main roof. Valley jack rafters are fitted either side of a valley rafter, as illustrated in Chapter 3, Fig. 3.10.

H – Dormer – the structure used to form a vertical window within a roof slope (see Chapter 3, Fig. 3.17 for other shapes of dormer). This structure gives increased floor area of full ceiling height within an attic roof construction, and is usually fitted with a window, hence the term 'dormer window'.

I – Barge board – the piece of planed timber is in fact a sloping fascia. It is often fitted to gable ends, as illustrated.

J – Dormer cheek – the term used to describe the triangular infill wall area between

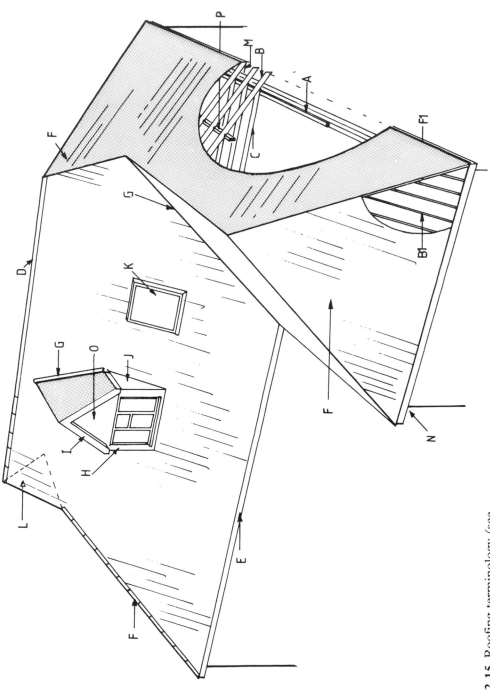

Fig. 2.15 Roofing terminology (see text for key).

dormer roof, main roof and the dormer front (see Chapter 3, Fig. 3.18 for the construction).

K – Roof window – sometimes termed roof light, the former being able to be opened for ventilation hence becoming a true window, the latter being fixed simply allowing additional light into the attic roof space.

L – Gablet – a small gable over a hip end. It is used as an architectural feature.

M – Soffit – the ply or other sheet material panel used to close off the space between the back of the fascia and the wall of the building.

N – Eaves – term used to describe the extreme lower end of the roof, i.e. the area around the fascia and soffit.

O – Gable – triangular area of wall used at the end of a roof to close off beneath the roof slopes. This is usually a continuation of the wall construction below.

P – Purlin – large section sawn solid structural timber, or fabricated beam, used to carry the common rafters on larger roof slopes where the commons are not strong enough or cannot be obtained in one single length, to span between the wall plate and the ridge (see Figs 3.2, 3.5 and 3.6).

CHAPTER 3
The 'Traditional' or 'Cut' Roof

DESIGN

The traditional or 'cut' roof as it has become known is essentially a roof cut and assembled on site from individual timber members. It is most frequently a common rafter and purlin roof, the design of which can be prepared from readily available standard span tables for the individual timber members. Hips and valleys are generally constructed to what has become known as 'good practice' and are less well documented with span tables and specific design aids. The sizing of these members is often left to the architect or engineer and it is not always necessary to provide calculations to prove their adequacy. The design of all new roof structures in England, Wales and Inner London must of course conform with the latest edition of the Building Regulations. In Scotland the Building Standards (Scotland) regulations apply, and in Northern Ireland the Building Regulations (Northern Ireland).

Whilst the Building Regulation statutory document provides the functional requirements with which one has to conform, the approved documents contain many span tables which, if used to design the individual timber members, will ensure compliance with the functional requirements and will be structurally sound. A guide to the timber members to be found in these tables and the limiting pitches is given in Fig. 3.1.

Other design aids can be obtained from TRADA, and on plywood applications for sheathing and stud gable end construction from APA – the Engineered Wood Association. Based on this information a roof may be designed and built for most common roof shapes.

THE COMMON RAFTER AND PURLIN ROOF

This simple form of roof is illustrated in Fig. 3.2. The structure is most commonly used where there is a gable at both ends of the roof, and is frequently to be found on terraced houses, as indicated in Fig. 1.10. Its construction has been included here because of the now very common refurbishment of such houses.

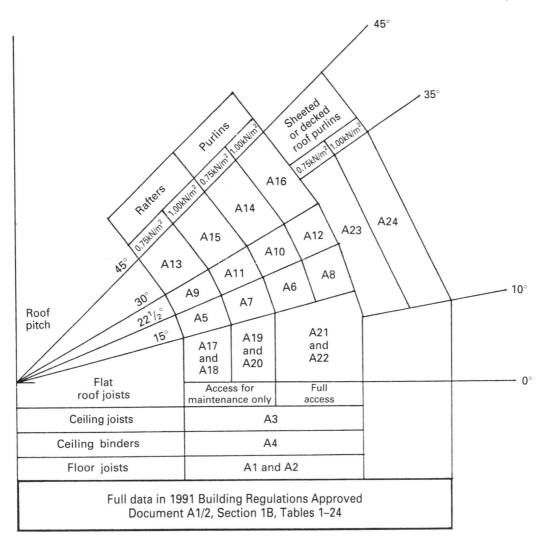

Fig. 3.1 Building Regulations table guide.

The wall plates are often simply bedded on mortar on either the inner skin of a cavity wall or, as is often the case with older terraced houses, on the inside edge of a solid 9 in. brick wall. Wall plates should be half lapped where they meet, and should not be less than 75 mm wide and 50 mm thick. They should be treated with preservative. Figure 3.3 shows typical plate connections. Further reference should be made to Chapter 7 where wall plates are dealt with in detail.

Purlins

In some of the older houses purlins were placed at right angles to the rafter. A more effective construction results with the purlins truly vertical for three reasons:

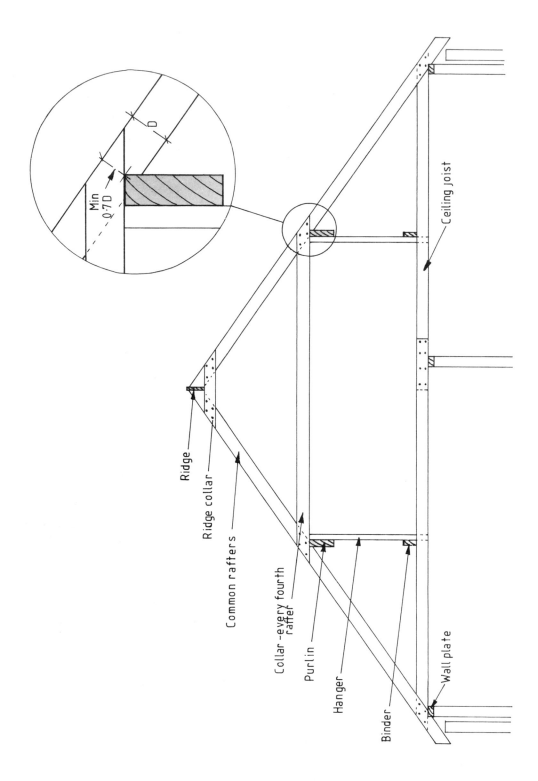

Fig. 3.2 Purlin and common rafter roof.

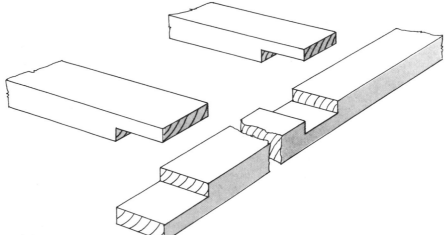

Fig. 3.3 Wall plate joints.

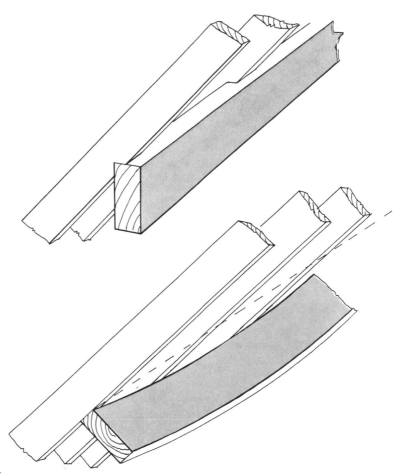

Fig. 3.4 Purlin deflection.

(1) The purlin is easier built-in or set in hangers at the gable walls.
(2) The purlin allows the rafter to be birdsmouthed over them, thus avoiding the tendency for the rafter to slide off the roof. A notch in the rafter can be used on sloping purlins but a birdsmouth is easier to locate and a quicker joint to cut on site.
(3) The sloping purlin has a tendency to sag down the roof slope thus necessitating a much thicker timber to maintain a true line. Figure 3.4 illustrates this point.

A common problem with this type of roof is the tendency to stretch the purlins structurally close to their design limit, so achieving maximum economy on the section of the purlin to be used. This sometimes can result in roof sag caused by deflection of the purlin, although the deflection may be within design tolerances. There are two ways of overcoming this problem. One is to design a stiffer purlin, i.e. probably one or two sizes up from the design table solution, the other is to stiffen the purlin using purlin struts, as illustrated in Fig. 3.5. The latter is to be preferred for, although slightly more labour intensive, it does allow ultimate economy in timber section and the struts give a stabilising effect to the walls supporting the purlins.

One final point on purlins: care should be taken with regard to the Fire Regulations when building purlins in to dividing or party walls between terraced dwellings. Unfortunately the approving authorities vary somewhat from area to area in their approach to timber built in to what are essentially fire walls between the dwellings, some allowing timber to be built in provided there is a positive fire break between the ends of the purlins, whilst others simply do not allow timber to be built in at all. In such cases built-in steel shoes will be necessary as indicated in Fig. 3.5.

On longer spans of purlins it may be necessary to use prefabricated beams, these being dealt with in more detail under the attic roof solution later in this chapter, the

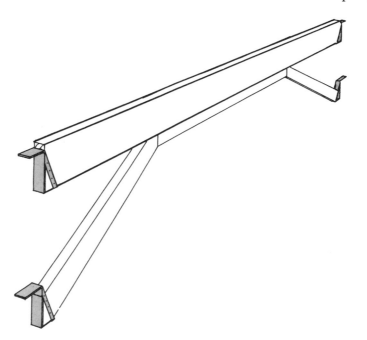

Fig. 3.5 Purlin struts.

beams themselves being illustrated in Fig. 3.13. An alternative solution on long spans is of course to provide an intermediate support for the purlin by means of a post which, in turn, is directly supported from a structural wall below.

Rafters

Little needs to be said about the common rafters as these can be simply designed from the span tables. However on a very long roof slope it may not be practical to obtain timbers in one continuous length. On roof slopes in excess of 4.8 m a second purlin should be considered as illustrated in Fig. 3.6. A collar should be fitted on every pair of rafters immediately beneath the ridge, and a further collar should be fitted to every other pair of rafters immediately above the purlin position.

Ceiling joist

Even on most domestic roof spans it will be impractical to obtain a ceiling joist member in one length. It will also be necessary, unless very large ceiling joists are used, to support the ceiling joist at some point along its length, by suspending it from the structure above and/or on a structural wall below. Figure 3.2 shows the typical solution using hangers and binders. More information on a typical hanger and binder combination for supporting the ceiling joists can be found in Chapter 4.

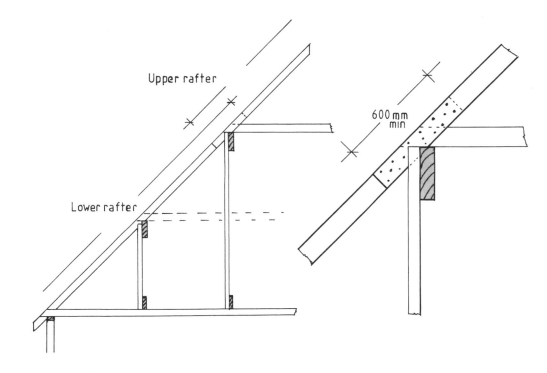

Fig. 3.6 Multiple purlins.

Connections

All timbers on this roof construction would normally be simply nailed together using 75 mm and 100 mm long galvanised wire nails. Whilst a simple 'tosh' or 'skew' nail (i.e. a nail driven at an angle through one piece of timber into its supporting timber) (see Fig. 8.6) will be adequate on the rafter to plate, purlin and ridge connection, the collar to rafter and ceiling tie to rafter connection should be made with three or five nails depending on the size of the individual members and taking care not to nail too near the ends of the timbers, thus avoiding splitting.

THE HIP ROOF

A simple hip roof is illustrated in Fig. 3.7. Whilst the wall plate is the main support for such a roof, the main problem of support arises from the lack of a gable end from which to support the purlin in the hip area. The mechanics of load distribution within the hip area seems to be open to debate. It is quite clear however, that the majority of the load is transmitted directly to the wall plate with the symmetry of the jack rafter leaning against each other either side of the hip rafter, tending to provide a self supporting structure. Certainly on small span roofs where no purlin is required this would be the case. On larger spans however, where a purlin is required on the longer jack rafters, a more sophisticated solution must be found.

The plate

The wall plate need only be a perfectly standard timber section, but with thrust from the hip rafter being resolved at the external angle of the wall plate, it is common to fit a tie across the corner. A more sophisticated corner joint used on some older buildings is illustrated in Fig. 3.8, but it would not generally be necessary for the size of the structure normally encountered on dwellings. It does however illustrate what was found necessary to contain the thrust from larger hip rafters.

The purlin

The need to support the purlin in the hip area has been mentioned above. One solution is to identify a suitable wall immediately beneath the hip area and use this for a support for the purlin. It is however more likely that the wall will be slightly outside the hip area as illustrated in Fig. 3.7, thus necessitating a degree of cantilevering of the purlin itself. The post should be at least twice the thickness of the purlin it supports, enabling a halved joint to be cut at the top thus allowing the purlin to fit squarely on a timber joint and not relying simply on nails. The supported purlin running across the hip end between the ends of the main purlins should again be halved onto the ends of the main purlin, thus providing a positive support.

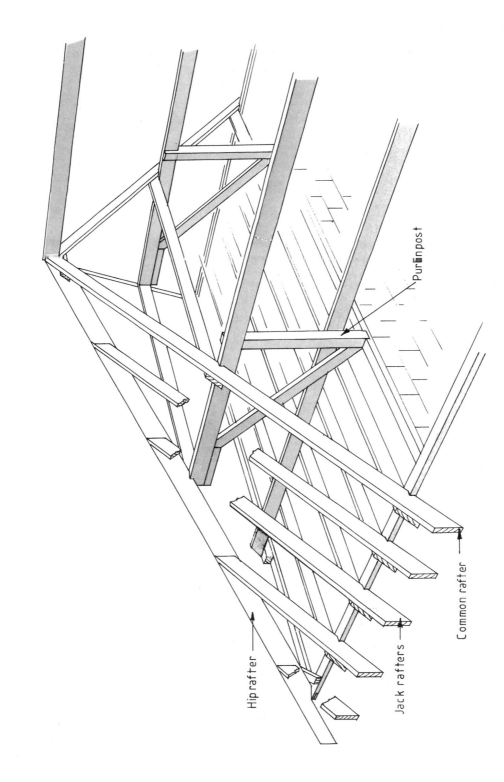

Purlinpost

Common rafter

Hip rafter

Jack rafters

Fig. 3.7 Cut roof hip construction.

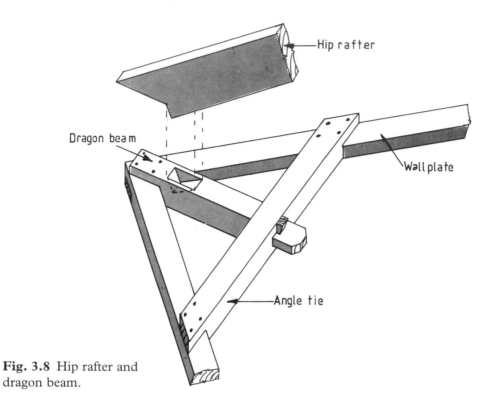

Fig. 3.8 Hip rafter and dragon beam.

Rafters

Rafters can be designed as for the more simple roof described earlier – the jack rafters will maintain the same cross-section, be birdsmouthed over the plate and nailed either side of the hip rafter. The angle of cut on the rafter abutting the hip rafter is what is known as a compound angle and this, like many other of the angles necessary on timbers in roof structures, can be calculated from the 'carpenter's square' or by reference to such specialist sets of tables as can be found in the *Roofing Ready Reckoner* (see the bibliography).

Ceiling joist

In the writer's opinion it is important to maintain the tying effect of the ceiling joist within the hip area and for this reason the ceiling joists should span in the direction indicated in Fig. 3.7. To do this it will be necessary to maintain the support by the binder within the hip area, and whilst the binder can be supported from the cantilevered purlin, it may be more prudent to have a separate binder beam supported at its ends, on the extreme end of the roof on the wall plate, and on an internal supporting partition wall.

THE VALLEY STRUCTURE

The valley is a very common feature of domestic roof structures. Some of the common roof shapes are illustrated in Fig. 3.9 with full valleys, i.e. a valley running from eaves to ridge in parts A and B, with shorter valleys in parts C and D.

If one considers the hip to be an external mitre of the roof, then the valley is an internal mitre. The easiest way to consider a structural solution is to imagine one roof being wholly or partially imposed upon the other, and this is most easily illustrated in Fig. 3.9B where the top part of the roof can be imagined to run through undisturbed as a normal gable to gable roof, with the leg of the T imposed upon it. Figures 3.9A–D show other valley situations.

Figure 3.10 shows a typical solution with valley jack rafters imposed upon a valley board which is itself supported by the main roof common rafters. The solution does of course assume that the common rafters will themselves be supported by a wall plate or beam immediately beneath the valley area. The alternative solution where this support is not provided is shown later in the solution for valleys on attic roof structures.

On valleys where the rafter length itself needs purlin support, the purlin in the valley area should be arranged at the same height as the purlin in the main roof. Furthermore it should be supported where it passes over the wall plate line of the main roof by a post, and by a steel hanger or shoe where it adjoins the main roof purlin. Figure 3.11 illustrates the purlin connections.

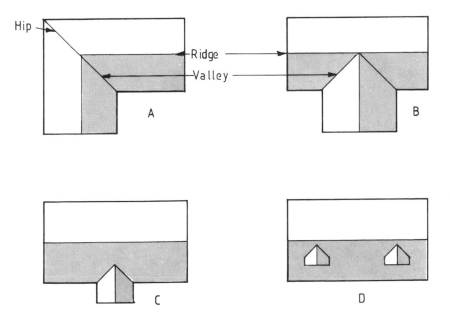

Fig. 3.9 Valley locations.

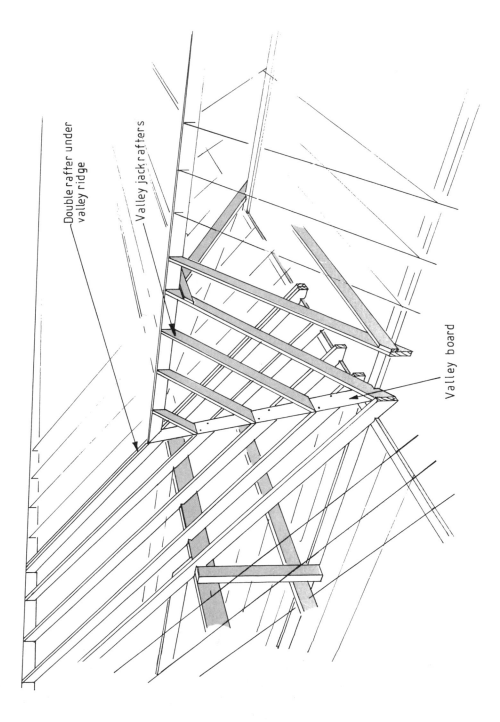

Double rafter under valley ridge

Valley jack rafters

Valley board

Fig. 3.10 Cut roof valley construction.

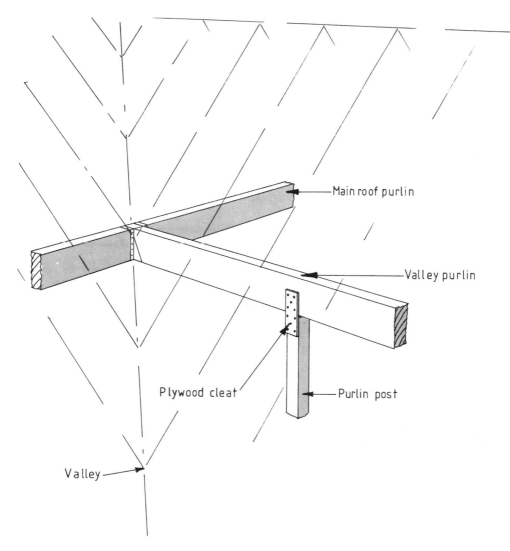

Fig. 3.11 Purlin support at valley.

ATTIC ROOFS

The attic or room-in-the-roof construction has become increasingly popular in recent years with the tendency for planners to seek steeper pitched roofs, particularly in rural areas. The house style created is often referred to as a 'chalet'. On a normal two-storey house with a roof pitch of about 45°, the volume enclosed by the roof is approximately 40% of the volume of the two storeys below, and on a single-storey building this proportion goes up to almost 80%. It therefore makes sense to attempt to use the extra enclosed space provided by the roof structure.

An attic roof structure with gables at both ends is comparatively simple to construct using purlins and common rafters; this roof shape can be seen in Chapter 7, Fig. 7.22. The hip end attic however poses some more difficult problems of support within the hip area, with the L- or T-shaped roofs posing further problems at the roof intersection. In all cases careful consideration must be given to the support both of the floor joists and of the purlins. The question marks in Fig. 3.12 show these problem points.

The simple attic

The floor joists will seldom be able to span from external wall to external wall and will therefore need some internal support either in the form of an internal wall, or a beam. Similarly the purlin is unlikely to be able to span from one gable to the other without internal support. Whilst the purlin problem can be eased by using beams capable of larger spans it is likely that some internal support will still be required. Examples of larger span beams are indicated in Fig. 3.13, giving spans of up to about 9 m without internal support.

Clearly it can be seen that in considering a common and purlin attic roof, the design must extend all the way down to the foundations below the load bearing internal walls. Support for the purlins by internal walls within the attic area will mean support in the form of beams or walls on the lower floor, thus to a certain extent controlling the room layout. A simple example of this is indicated in Fig. 3.14, based on the layout of Fig. 3.16.

It is therefore almost impossible to provide a solution to an attic construction without knowing precisely the room layout, and which of those room walls is capable of carrying load. To aid construction on site, the design should have:

(1) Common floor joist depth;
(2) Common rafter depth;
(3) Common purlin depth;
(4) Common purlin lines.

Structural economy can be achieved by varying the thickness of individual members and/or their grade stress, and the spacing of structural members.

The hip and valley attic

Having highlighted the need for both floor joist and purlin support, on a simple attic, consideration must now be given to the solution of the more complicated attic roof as illustrated in Fig. 3.12.

It can be assumed that the floor joists will generally be supported either by beams or by load bearing walls below. In some instances these joists may be doubled to form a beam, or a separate beam is inserted within the floor which will itself be strong enough not only to carry the floor but also to support the purlin above. This however generally applies only to the lower purlin, i.e. the one at the attic wall to the sloping ceiling

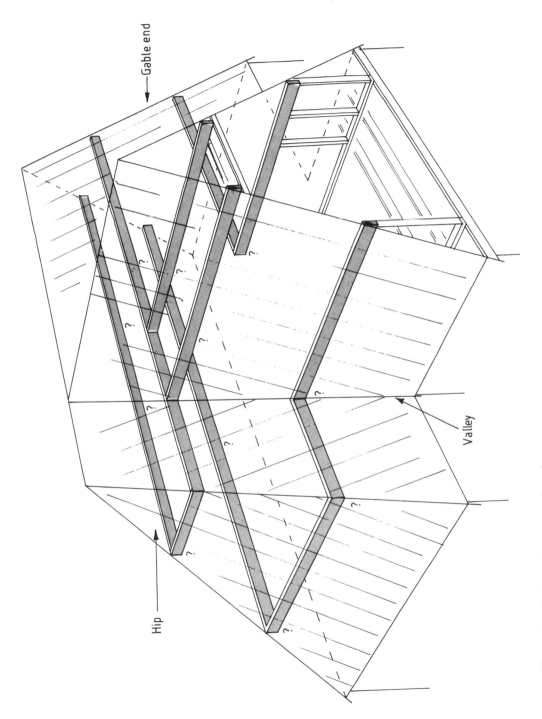

Fig. 3.12 Hip and valley purlin support locations.

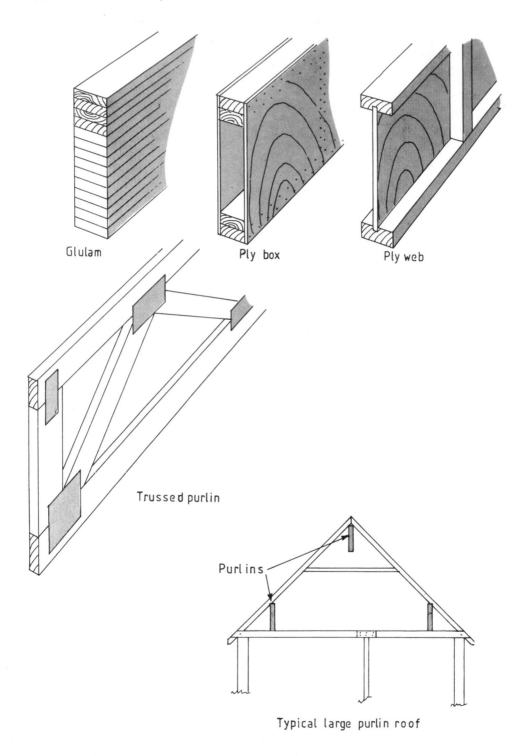

Fig. 3.13 Alternative purlin beam constructions.

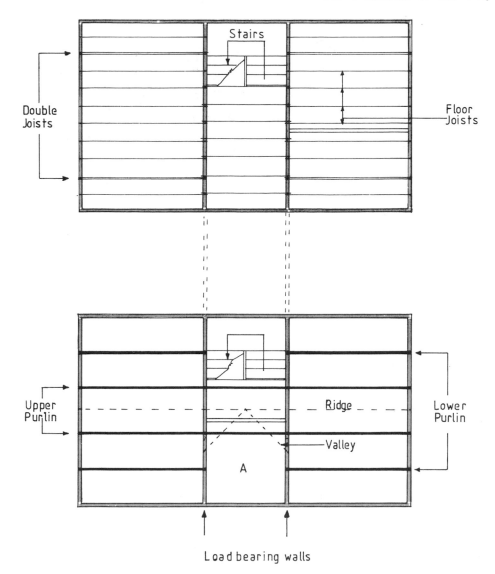

Fig. 3.14 Purlin support – attic roof.

junction, and not to the upper purlin. The latter is more likely to be supported by internal load bearing walls.

Reference should be made to Fig. 3.15. At the gable end both sets of purlins can simply be built-in in the usual manner, with the floor joists supported by being built-in at the gable and either on a load bearing room dividing wall internally or a beam as illustrated.

At the hip end the construction is more complicated, but again floor joists can be supported on the wall plate of the external wall and on either a beam or internal load bearing partition. The lower purlin at the end of the hip can be either supported at its

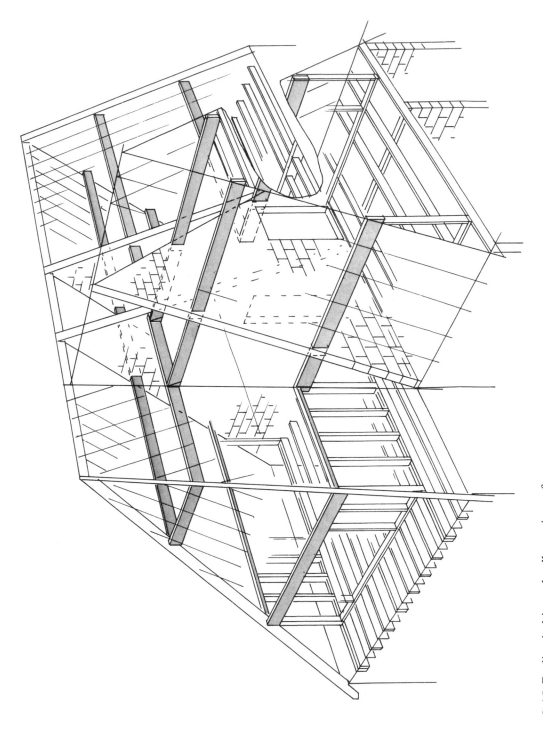

Fig. 3.15 Purlins in hip and valley attic roof.

ends, strutted up from a specifically designed double joist or beam below, or it can become little more than a wall plate supported by the end wall studs from floor joists below. In the latter situation of course the floor joists must be designed to carry this additional load. The side purlins in the hip area are most likely to become in effect wall plates, themselves being supported by studs at 400 mm or 600 mm centres, the studs being supported by the double joist or beam below. If the side wall is to be omitted to give maximum floor space then the purlin must be capable of spanning the room width and will need to be supported by room dividing walls.

The upper purlin will generally span from gable to internal load bearing walls and then from internal load bearing walls to further load bearing internal walls as illustrated. However in the hip area it is not possible to provide an adequate cantilever length, bearing in mind minimum room size requirements, and it is certainly not possible to provide a strut as illustrated in Fig. 3.7. The purlin at the hip rafter junction therefore may well have to be supported by the hip rafter itself, thus making the hip rafter a major structural element. Where this is the case it will not be possible to design the roof structurally from Fig. 3.1 and design advice must be sought. For the purposes of the illustration such a structural rafter has been assumed.

The valley intersection
The lower purlin in Fig. 3.15 is supported by the internal load bearing wall and will to a certain extent cantilever into the valley area. However additional support may be provided by a post down onto the beam provided in the floor of the main roof structure. The upper purlin again supported by the internal load bearing wall will not be able to provide a full cantilever and must therefore be connected into the purlin within the main roof structure. The connection will normally be by a steel shoe and is of course following a similar structural layout to that indicated in Fig. 3.11, with the purlin post being replaced by the internal wall. Rafters, collars and ridge may be provided bearing in mind the considerations described earlier, especially concerning rafter length.

Attic dormers

Most attic roofs will be fitted with at least one dormer to provide both increased full-height room area within the roof, and light and ventilation. Roof windows may be fitted to provide light and ventilation but they do not add significantly to the full-height room area.

The first dormer type construction to be considered is that illustrated in Fig. 3.16, the main structure of which can be seen to be brickwork continued up from the structure below. Such a structure would provide the roof over the area marked A in the lower illustration on Fig. 3.14. It can be seen that this follows a conventional valley situation, with valley jack rafters supported by a ridge onto a valley board in turn supported by the common rafters of the main roof. The common rafters in this case, rather than being supported by a wall plate, are supported by the upper purlin of the attic roof. The true dormer window, examples of which are illustrated in Fig. 3.17, occurs within the roof slope, i.e. not starting at the eaves and probably finishing well before the main roof ridge. Figure 3.9D shows two such dormers.

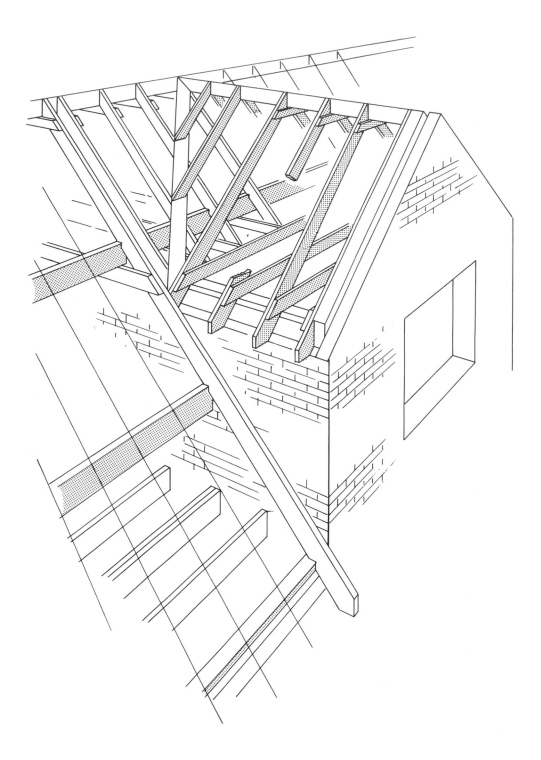

Fig. 3.16 Purlin support on internal walls.

Fig. 3.17 Dormer styles.

The additional load imposed by the dormer on the main roof is relatively small, bearing in mind that the most significant weight on a roof is the tiles, and this tile area of course is not increased. The additional load must be allowed for.

Dormers, depending on the architecture of the area in which the house is to be built, will have a variety of roof shapes. Whilst Fig. 3.17 illustrates a flat, mono-pitch or 'catslide' and a conventional symmetrically pitched roof with gable ends, hip end dormers are not uncommon.

Dormer framework

The construction of the dormer is relatively simple. The critical part of the design is the forming of an adequately strong framed opening within the main roof. In forming the hole for the dormer, the continuity of roof slope support is removed and provision must be made to carry the load both above and below the opening formed. If an upper and lower purlin are used in the attic structure, then these members may be used to support the rafters both above and below the opening. The perimeter of the hole formed in the main roof will provide the foundation for the dormer framework itself. Figure 3.18a illustrates the method of imposing simple dormer framework onto a trimmed opening. Figure 3.18b illustrates the rules for trimmer numbers.

ROOF LIGHTS AND ROOF WINDOWS

Roof lights generally will require much smaller openings within the main roof structure than a dormer described above, therefore a similar method to that illustrated in Fig. 3.18 will be more than adequate. However, as the roof lights may not extend up the roof slope the full distance between two purlins, separate secondary purlins or trimmers may have to be introduced. If this is the case the rafters onto which the trimmers are fixed must be reinforced by attaching an additional rafter to each side of the opening, these additional rafters extending from the lower to the upper purlin (see Fig. 7.26).

The term 'roof window' is a term recently introduced to the building industry to describe what is in effect an opening roof light. The roof light is normally fixed, of course, and provides only light to the building. Roof windows are most commonly of proprietary manufacture with the manufacturers providing detailed guidance on the method of fitting the roof window to both existing and new roof structures. Reference should be made to the manufacturer's instructions if such a roof window is to be fitted.

The advantage of the roof window over the dormer, if additional floor space is not the criterion, is that because the glazed area is angled directly at the sky, significantly more light is admitted to the room. The manufacturers of the proprietary roof windows also claim up to 70% saving in cost over a comparable dormer construction.

ADDITIONAL DESIGN CONSIDERATIONS

New houses constructed under the control of the *National House-building Council* must further conform to the requirements of the *Registered House-Builder's Handbook*, Volume

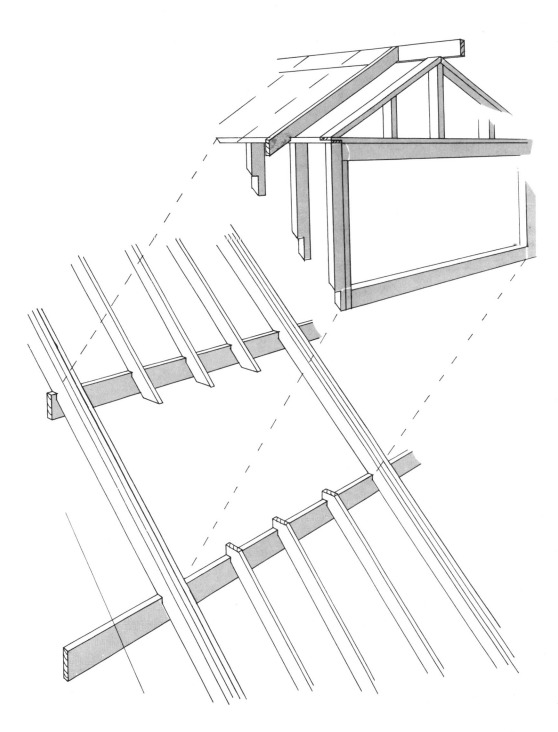

Fig 3.18a Trimmed opening for dormer. For trimmer numbers required see Fig. 3.18b.

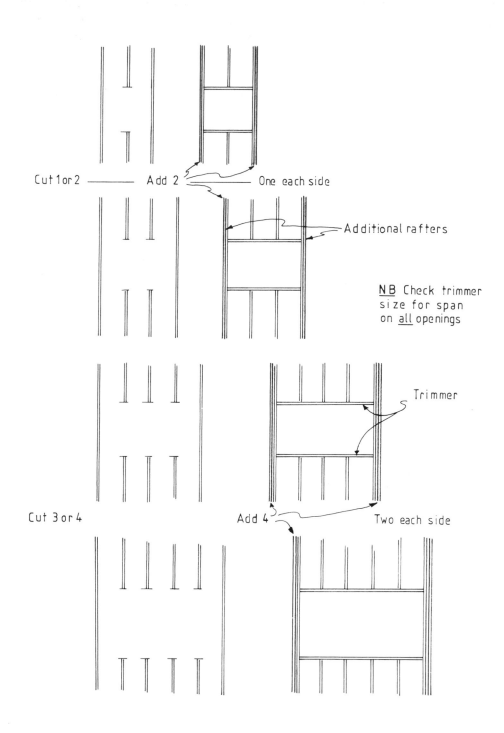

Fig. 3.18b Trimming construction rules.

2 of which sets out the technical requirements for the design and construction of dwellings. The reader is directed particularly to the section on carpentry, Chapter 7.2.

Chapter 7 of this book concerns itself with the many other aspects of roof construction which apply to all roof forms. The items covered are as follows:

(1) Storage and handling of timber;
(2) Preservative treatment;
(3) Wall plates and straps;
(4) Gable ends, straps and gable ladders;
(5) Water tank platforms;
(6) Ventilation of roof voids;
(7) Roof bracing;
(8) Eaves details;
(9) Trimming small openings.

This chapter has dealt with the construction of new attic roofs. Chapters 8–11 explore the possibilities, problems and solutions when converting loft spaces to form attic rooms.

CHAPTER 4
Bolted Truss Roof Construction

Roofs of all shapes and sizes can be structurally designed using bolt and connector jointed truss forms. Individually engineered designs for bolted trusses are used for public buildings where they are often left exposed as a feature of the design of the building. The bolted truss designs most frequently found in domestic dwellings are those prepared by TRADA, and it is these designs which will be dealt with in this chapter.

The bolted trusses to be found in many houses built prior to the trussed rafter era were based on designs by TRADA. Regrettably these are no longer available in the new stress grades in existence today. However, these designs may still be copied when extending a roof, converting to current timber grades and will continue to be a useful form of DIY roof construction.

THE JOINTS

The bolt and connector roof structure is comprised of principal trusses spaced at centres dictated by the structural design. The principals in turn support purlins and common rafters, with binders supported by the principals, carrying the ceiling joists. The strength of the principal is quite naturally in its joints between the timber members, and therefore careful assembly of the principal truss is essential if the roof is to perform satisfactorily. Many such roof forms have been constructed on dwellings since the advent of the TRADA designs, and it is the sturdiness of the design, rather than high standards of workmanship, which has led to the method's success.

A simple bolt and connector joint is to be seen in Fig. 1.11. This illustrates a double-sided toothed plate connector which is held in position (having first been embedded in the meeting timber surfaces) by a bolt with large load spreading washers. The alternative to the double-sided toothed plate, is the 'split-ring' connector and this will be dealt with later. It is essential that the teeth of the timber connector be fully bedded into the timber surfaces if full design strength is to be developed. It is the connector and not the bolt which is the principal strength of the joint, the bolt mainly being there to hold pieces of timber together and ensure embedment of the connector is maintained. It is therefore essential that the nuts on the bolts are tight and some retightening may be necessary once the truss form has been installed as timber shrinkage may have occurred. It is quite common to have two or more connectors at a joint and the connectors themselves may vary in size and shape depending upon the design specification.

TRUSS ASSEMBLY

To understand the assembly sequence about to be described, reference should be made to Fig. 4.3 which illustrates a typical connector jointed truss. For the purposes of the assembly sequence it will be assumed that all joints are made with double-sided toothed plate connectors.

Timber for one truss should be selected, cut to size, laid out on a flat surface into the precise truss shape required and all joints clamped together. Dimensions and pitch angle must now be checked and, if all are found correct, one can proceed to mark out the bolt centres and drill with a bit diameter not greater than 1.5 mm larger than the bolt diameter specified in the design. Care must be taken to maintain the drill square in all directions with the timber surface. If more than one truss assembly of the shape laid out is to be constructed, then this first truss must now be unclamped and the components used as 'masters' for the members of the remaining trusses to be constructed – this will not only save labour but will ensure that all of the trusses are identical. At this stage, then, the masters should be used to cut and drill all of the members required for the remaining trusses.

To assemble the first truss, lay out all the members and place a bolt through all of the joints except the first joint to be fitted with a connector. At this first joint (probably the rafter to ceiling tie joint) fit the specified connector and, using a special high tensile steel stud available from the connector suppliers, pass this through the bolt hole in the joint and place a large 100 mm square × 6 mm thick steel spreader washer on both sides of the joint. By turning the nut, bed the connector fully home. Considerable pressure is needed to bed the double-sided tooth plate timber connector, and the mild steel bolts used to maintain the joint in the final assembly are not adequate for three reasons: firstly, the pressure required may well strip the threads from the mild steel bolts; secondly, if several connectors occur on the same bolt line there is unlikely to be adequate length of thread on a standard bolt to pull the joint down embedding all of the connectors (Fig. 4.1 illustrates this). Thirdly, the smaller 50 mm × 50 mm × 3 mm thick mild steel washer used with the standard bolt will be inadequate to protect the timber surface from crushing and may itself be severely distorted. Where two or more sets of connectors occur, such as the ceiling tie to ridge strut joint on Fig. 4.3, the appropriate number of high tensile studs will be required to pull the timbers down progressively if severe distortion of the timbers is to be avoided, as illustrated in Fig. 4.1.

When the connectors are fully embedded the high tensile stud can be withdrawn leaving the joint generally well held together with the connectors. The standard bolt may now be inserted in the joint with the 50 mm square washers under both head and

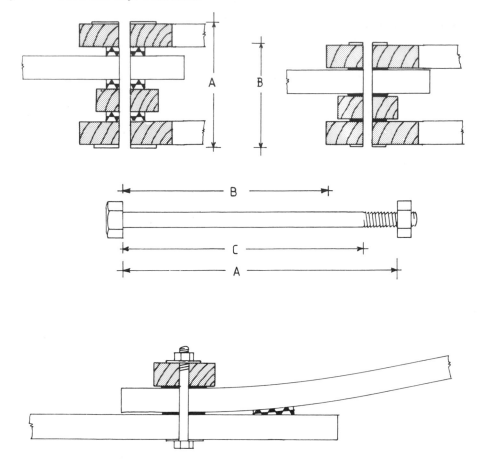

Fig. 4.1 Toothed plate connector joints.

nut. The only pressure now required is that to maintain the joint in its bedded position. At this stage the nut should not be unduly tightened.

The TRADA designs (referred to in Chapter 1) showed no collar under the ridge board, consequently for handling at the works (if the trusses are made off site), during transportation and also for on-site handling the truss is not jointed at the ridge. A temporary collar should therefore be fitted to avoid distortion of the truss and this collar should now be nailed into position on the rafters just below the ridge board line. On site, this will then aid the location of the ridge board during the roof construction. The space in the setting out of the truss to allow for the ridge board must not be varied because clearly it is vital to the overall geometry of the truss. A thinner ridge board will mean that the truss will sag and a thicker will result in the reverse.

The remainder of the joints in the truss can now be assembled as described above and when all are complete all the bolts should finally be tightened before the truss is moved. Subsequent trusses can now be assembled using the masters following the procedure described.

THE SPLIT-RING CONNECTOR

The split-ring connector joint is illustrated in Fig. 4.2. The principle is similar to that in the double-sided tooth plate, in that the split-ring carries 75% of the load, the bolt 25% and maintains the timbers in position around the ring. The problem of embedment of the connector is solved by machining a circular groove equal to half of the width of the ring in each meeting timber surface. This groove is machined using a special 'dapping' tool available from the connector suppliers. Assembly of the split-ring connector jointed truss is similar to that for the design described above, but clearly all of the grooves for the rings must be machined before any final assembly takes place. The high tensile stud is clearly not required as the ring fits neatly into the groove allowing the final bolt to be placed through on the first assembly. The split-ring connector truss assembly is most likely to be carried out in a workshop where the dapping procedure can be carried out in fixed drilling jigs on benches. Dapping can be carried out on site but electric power to turn the dapper is almost essential, and a degree of skill is required in its accurate use. The bolt holes must be drilled first; the dapping tool is then located in the bolt hole to cut the groove for the connector.

　　Site assembly of bolt and connector jointed trusses using either connector form is of course a practical proposition, but factory assembly of the truss will generally result in a

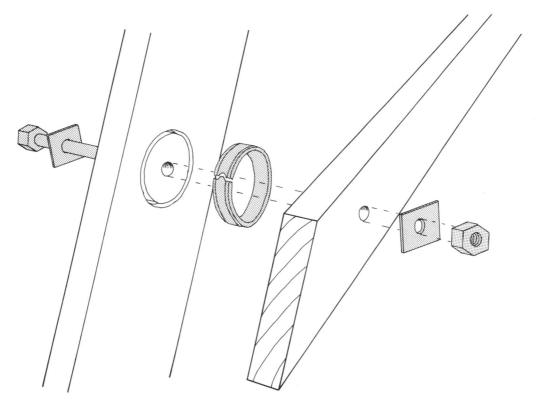

Fig. 4.2 Split-ring connector joint.

more accurate component simply because of the equipment, level floor area, dry conditions and specialist operatives used in the construction.

ACCURACY

Most bolt and connector trusses for domestic roof construction will be assembled using sawn timber. The tolerance on sawn timber in terms of over and under sizing, particularly in the larger dimension, is significant and for this reason the full size setting-out of the truss has been recommended in the assembly procedures described. The precise location for the bolt centres is essential to maintain the design edge and end distance for the timber connectors themselves. Insufficient end distance may result in a timber connector shearing out a section of timber thus allowing the truss joint to fail. Great care must therefore be taken to follow the precise dimensions detailed on the engineer's drawing. Reference should be made in this respect to the dimensions indicated on the design illustrated in Fig. 4.3.

STANDARD DESIGNS

The TRADA designs as stated earlier are no longer available, but the design principles are sound and all relevant metalwork and connectors are still available. Figure 4.3 is continued in this third edition by way of illustrating this typical bolt and connector truss but it must be emphasised that the illustration is *not* a current TRADA design.

Many homes built in the fifties, sixties and seventies used this type of construction and it has proved to be a rugged durable roof structure. Similar roof trusses today would have to be engineered, but that done, purlins, common rafters and ceiling joists can all be 'designed' from building regulation span tables.

THE ROOF CONSTRUCTION

The bolt and connector truss is generally used as a principal truss with common rafters and ceiling ties supported from it and on the wall plate. The designs referred to above are spaced at 1.8 m centres with the design of the purlin, binders, plate and ridge being given on each individual sheet. A study should be made of the design sheet illustrated in Fig. 4.3.

The construction of the roof itself, once the principal trusses have been produced, is quite straightforward, with many of the joints being nailed as with the 'traditional' roof. The exception to this in some designs is the joint between ceiling joist and rafter, and on all designs where the ceiling tie cannot be obtained in one length, also the splice joint between the two lengths, which again uses a combination of connectors and nails. An alternative for the ceiling joist allows for it to be joined over a plate on a partition; this is not to be recommended where the construction process requires non-load bearing

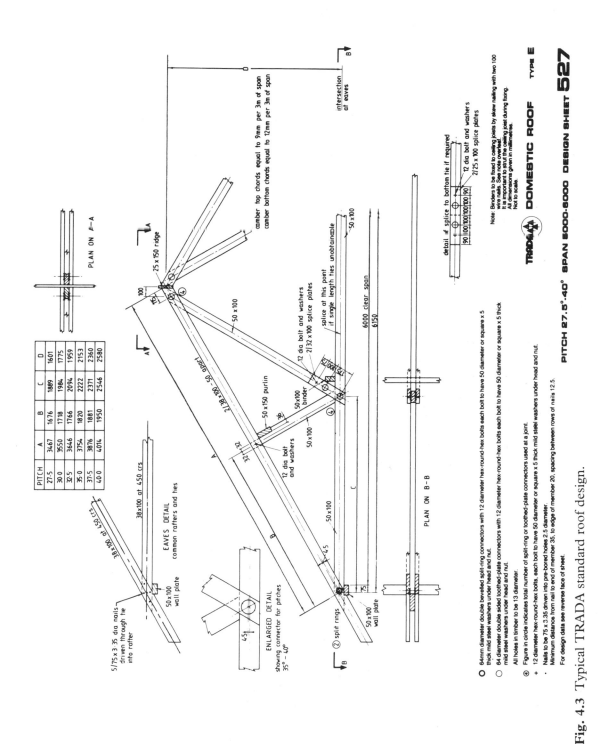

Fig. 4.3 Typical TRADA standard roof design.

partitions to be installed after the roof is constructed. The roof should therefore be constructed wherever possible as an independent clear spanning structure between the two wall plates of the external walls. To avoid trusses inadvertently bearing on internal non-load bearing partitions, it is good practice to have the roof tiled and therefore under normal working load conditions to allow any deflection in the truss to take place before the partitions are fitted.

The trusses are not designed to carry water storage tank loads and thus, wherever possible, these should be directly supported from partitions below. If this is not practical then the advice of TRADA or the truss designer must be sought in order that the truss spacing may be reduced to carry the additional load. One final point on the standard roof assembly is that some difficulty may be experienced in nailing the ceiling joist tightly to its binder. A more effective connection is to use one of the readily available light galvanised metal cleats. These give a much more positive and stronger connection than traditional skew nailing. Figure 4.4 illustrates the roof construction described above.

BOLTED TRUSS HIPS

Hips can be formed using bolt and connector trussed structures, the hip usually being supported on a 'half' truss, itself supported on one of the principals at hip peak and on the end wall plate at the hip eaves. A typical hip construction is illustrated in Fig. 4.5. The remainder of the hip timber members follow basically traditional construction techniques, described in Chapter 3. The hip rafter, however, generally carries the ends of the purlins and also supports the hanger, which in turn helps to span the binder from first principal to the end wall plate. It is essential that the ceiling tie member of the half truss used in the hip end is connected properly into the binder running on the centre line of the roof to ensure an adequate tie from one end of the roof to the other.

VALLEYS

Valley construction follows the design set out in Fig. 3.10, a principal truss being located on the ridge line of the intersecting roof in place of the double rafter illustrated. With equal span and pitch intersecting roofs, purlin lines will coincide and they can be fixed together with a steel connector as indicated in Fig. 3.11. Unequal spans will mean a hanger from the higher to lower purlins and/or additional support again as indicated in Fig. 3.11. Support may also be required for the main roof principal truss if there is no wall on the normal plate line. It is not generally possible to support this principal from the first principal truss of the intersecting roof unless it has been specifically designed to carry the additional load. It is more likely that a steel or timber beam must be employed within the roof space, using a special steel shoe designed to carry the end of the principal truss. Figure 4.6 illustrates this junction. A beam beneath the truss is a more easily constructed detail, but its depth may restrict headroom below.

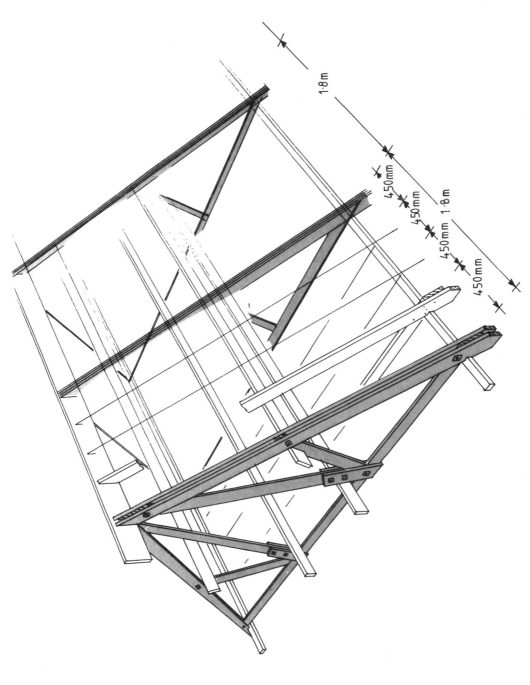

Fig. 4.4 Bolted roof truss construction.

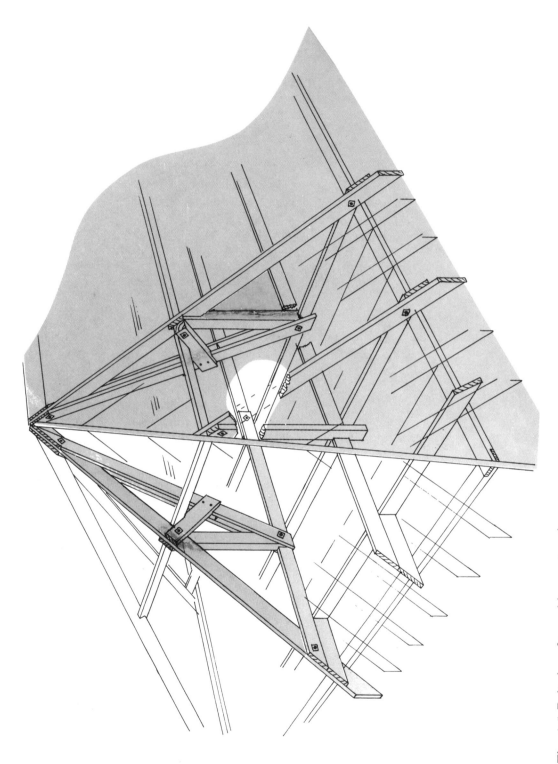

Fig. 4.5 Bolted roof truss hip construction.

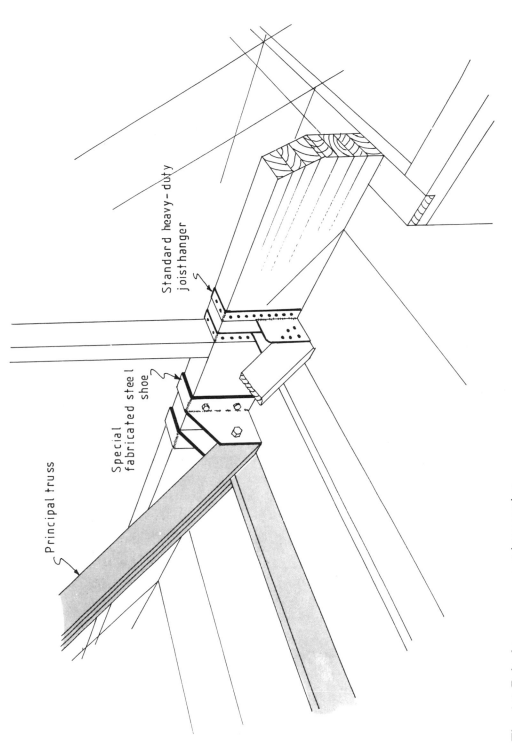

Standard heavy-duty joist hanger

Special fabricated steel shoe

Principal truss

Fig. 4.6 Bolted truss support at intersection.

Care must be taken with this steel shoe design and the positioning of the supporting beam, to allow adequate 'end-distance' for the rafter to ceiling joist connector discussed earlier. Also the positioning of the truss to shoe locating bolt should ideally be identical to the bolts used to assemble that particular joint. Assembly on site would then require temporary removal of one of the joint retaining bolts whilst the truss is temporarily supported. A slightly longer bolt is passed through the steel shoe and the truss to complete the joint.

Lighter weight steel standard hangers can be used to support rafter and ceiling joist, although the location of the supporting beam may be such that an extended length of support in the shoe similar to that used for trussed hangers may be required to give adequate bearing to the supported structure members.

STRUCTURAL OPENINGS

Dormer winders are unlikely to occur in a bolt and connector roof unless a specific attic design has been produced, in which case all openings will also have been structurally designed and these must be followed. Openings for chimneys and roof hatches will occur and these can generally be dealt with as illustrated in Fig. 3.18 or, in the case of small openings, as illustrated in Fig. 7.26. Care must be taken to position the principal trusses at design stage to avoid such openings.

ROOF STABILITY

The bolt and connector jointed principal roof truss construction, like most other forms, should not rely for its lateral stability upon the gable end wall structure. For this reason diagonal bracing on the undersides of the rafters should be provided and of course some temporary bracing will be required at construction stage to maintain the heavy principal trusses in position. Roof bracing is dealt with in more detail in Chapter 7.

CHAPTER 5

The Construction of Trussed Rafter Roofs

The majority of roofs constructed for domestic dwellings in the United Kingdom now use the punched metal plated trussed rafter construction. Over two million units are produced each year. The majority of the trussed rafters are produced in factories where capacity to both design and produce varies enormously – from those manufacturers able to design the trussed rafters only, to those fully competent timber engineering companies employing their own structural designers. Production capacity ranges from 200 to 5000 trussed rafters per week, with quality also varying both in the final product and in the service offered to the customer. The punched metal nail plate connector can only be fixed using the specialist equipment needed for pressing the plates into both sides of the timber joint. It is *not* possible to fix them on site. Figure 5.1a illustrates a typical punched metal plate joint.

There are two alternative systems to the punched metal plate fastener, one being a metal plate punched with holes which is then fixed to the timber joint with special twisted nails (Fig. 5.1b). The other is to use plywood gusseted joints, the plywood being fixed either by glue with nails to hold it in position whilst the glue cures, or exclusively by nails of designed size and fixed to a specific designed pattern on the joint: Fig. 5.1c illustrates such a joint. With the exception of the glued option, the latter two methods are suitable for site assembly.

PERFORMANCE IN USE

Much has been written in trade journals concerning possible problems occurring with the trussed rafter form of construction, with reference particularly to its long-term durability. An authoritative paper was prepared in 1983 by the Building Research Establishment entitled *Trussed rafter roofs* (IP14/83) in which the results of a nationwide survey were summarised. The survey looked at the manufacture, site use, performance in service, and plate corrosion with certain types of preservative. Whilst some short-comings were found in the manufacture, the areas causing greatest concern were site

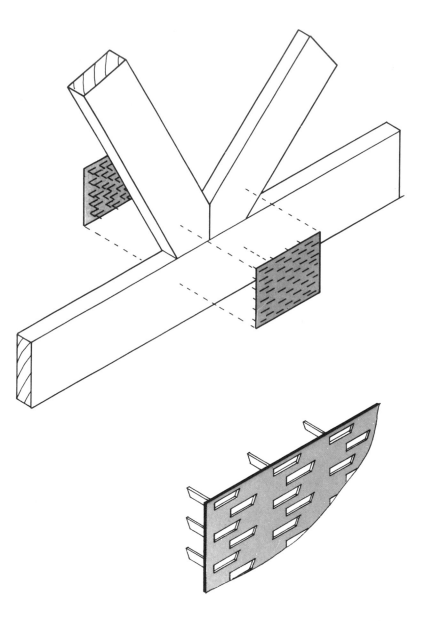

Fig. 5.1a Punched metal plated joint.

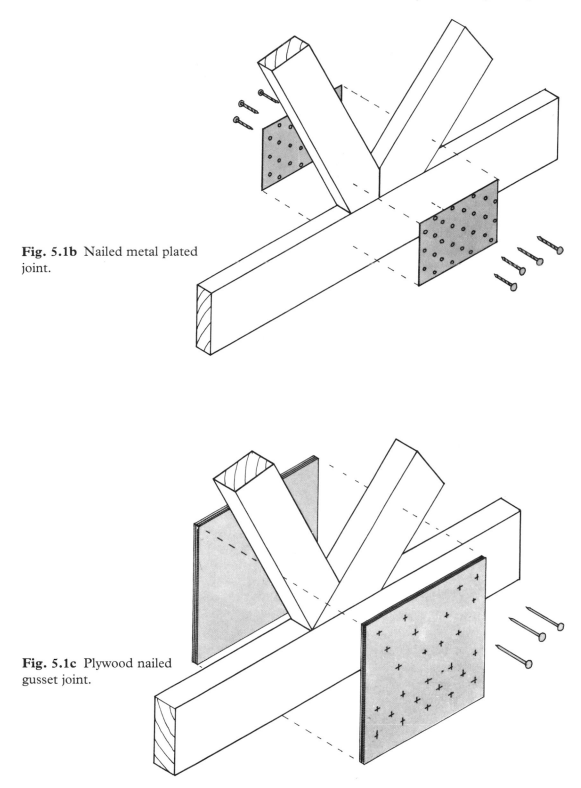

Fig. 5.1b Nailed metal plated joint.

Fig. 5.1c Plywood nailed gusset joint.

handling and construction of the roof structure using the trussed rafters on site. Inspection of the trussed rafter components for compliance with the relevant British Standards before assembly into the roof structure should be made, thus overcoming the possibility of faulty structural components being installed. The problem of inadequate design and assembly information for the roof structure as a whole remains, and whilst BS 5268: Part 3 gives some guidance for conventional gable ended roofs and the individual plate manufacturers provide standard details for hips and valleys, the builder is not generally presented with a specific set of drawings for the assembly of the roof on which he is working. Increasingly the larger trussed rafter producers and the plate manufacturers are providing computer programs which generate 'whole roof' designs. These designs not only include the layout of the trussed rafter itself, but all other infill timbers and support metalwork required for the structure. This considerably aids the correct site assembly of the roof structure.

This chapter concentrates on the construction of trussed rafter roofs including hips, valleys, attics and trimmed openings. The storage and handling of trussed rafters and other timber components, and the many ancillary details required to complete the roof are set out in Chapter 7.

DESIGN

The structural design, manufacture and some aspects of construction are dealt with in BS 5268: Part 3: 1998, *Code of practice for trussed rafter roofs* and reference will be made to this important document throughout this chapter. Whilst timber sizes may be obtained from safe span tables, the design of the joint plate or gusset *must* be provided by a qualified structural engineer.

Most trussed rafter manufacturers use terminology not yet covered in previous chapters and the reader should refer to Fig. 5.2 for familiarisation with terms to be used throughout this chapter. The illustration shows a 'fink' truss, the most common configuration in use today. The geometry of this configuration can be found in Fig. 2.5. The standard spacing for trussed rafters is 600 mm, although 400 mm and 450 mm are not uncommon. Timber sizes are standardised throughout the United Kingdom, the timber being machined on all surfaces for accuracy in accordance with the standards set out in BS EN 1313/1. The timber is usually stress graded in accordance with BS 4978 and should be stamped with a grade mark. Whilst so called 'nominal' sizes are often quoted, i.e. 75, 87, 100, 125 and 150 × 38 mm, the finished section will be 72, 84, 97, 122 and 147 × 35 mm. Timber of 47 mm finished thickness is frequently used for attic trussed rafters with depths going up to 222 mm for heavily loaded rafters and floor joists. Trussed rafters in excess of 11 m span must use this thicker timber, or be made of multiple trusses of minimum 35 mm thickness, permanently fixed together by the truss manufacturer at works.

An understanding of the function of the trussed rafter is essential if good roof construction is to be achieved. The trussed rafter is designed to carry only the vertical loads imposed upon it, no lateral loads are catered for. The design assumes that the trussed rafter is maintained in its truly upright position by the various bracing and

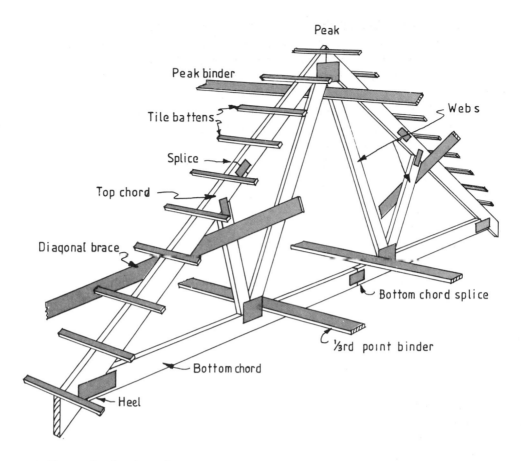

Peak

Peak binder

Tile battens

Splice

Top chord

Diagonal brace

Webs

Bottom chord splice

⅓rd point binder

Bottom chord

Heel

Fig. 5.2 Trussed rafter lateral support.

restraining members. Figure 5.2 illustrates this point. The wall plate, binders, tile battens and diagonal braces are all assumed in the structural design. All of these items have to be site fixed and it is essential therefore that the specification for these fixings is carried out. The load on the top chord is from self weight of the truss, the tiles, felt and battens plus statutory snow loading. On the bottom chord the load is from the self weight, ceiling and insulation plus a nominal loft loading and an additional load imposed by a man walking on the bottom chord. No other loads are designed for. Wall plates must be a minimum of 75 mm in width (unless structural design specifies otherwise), to avoid the load from the truss crushing the timber on the underside of the bottom chord or the wall plate at the point of contact. As a guide the bearing length should not be less than 0.008 times the span of the trussed rafter. This effectively means that on spans above 9.3 m it is essential to use 100 mm wide wall plates and this indeed is the most common practice, this width matching precisely the width of the inner skin of a conventional cavity wall construction. Refer to section 7.3 of BS 5268: Part 3 for more detailed information.

DESIGN INFORMATION

It can be seen that, when requesting a quotation and particularly when ordering the trusses, it is essential to inform the trussed rafter manufacturer of not only the truss shape and size and number required, but also the roof tile type and ceiling specification. The best way to ensure that the design is adequate for the roof under consideration is to send a full set of drawings for the building project such that this, or an additional specification, includes the information set out below.

Information required	Reason for data requirement
(1) Dimension between walls	To obtain accurate span over wall plates
(2) Size of wall plate	To obtain accurate span over wall plates
(3) Location of wall plates or other supports for the truss	To obtain accurate span over wall plates
(4) The overall thickness of the cavity wall	To determine the top chord overhang required
(5) Width of soffit required	To determine the top chord overhang required
(6) Pitch in degrees, left hand side of roof slope	To establish the geometry of the roof truss
(7) Pitch on right hand side of roof slope	To establish the geometry of the roof truss
(8) Any minimum top chord size requirement, particularly if these are to match existing construction.	Structural design may produce a rafter of smaller section if no limit placed upon the designer
(9) The size and position of all water tanks and other equipment or load supported by the trussed rafters.	For design load calculations
(10) Size and location of loft hatch and any other openings such as staircases in an attic construction.	To design any additional trusses or trimmings for openings
(11) Size and location of any chimney stacks	To design any additional trusses or trimmings for openings
(12) Roof tile or other covering specification giving manufacturer's name and type or a precise weight	For loading calculations
(13) Ceiling and insulation specification	For loading calculations
(14) Timber preservation specification requirement	Effective timber preservation must be carried out after the components are cut and before assembly
(15) Overall length of the building	To allow the correct number of roof trusses to be calculated
(16) Truss spacing required	To allow the correct number of roof trusses to be calculated

(17) Site address and location of building:	To inform designer of wind and snow loads on building and to aid delivery of goods to site
(a) The height, ground roughness and location of the building making reference to any unusual wind conditions that may exist	
(b) The site snow load if known or the basic snow load and altitude of the site, or the Ordnance Survey grid reference for the site	
(18) Proposed use of the building with reference to any unusual environmental conditions	This would cover such items as a swimming pool or a canteen, thus giving the roof designer an insight into the likely conditions in which the roof has to work
(19) The position, size and shape of any adjacent structure which may be higher than the proposed roof and any building which is closer than 1.5 m	This is to give information for both wind and snow load calculations
(20) Any special requirement for the minimum member thickness	There may be a minimum thickness for fixing either a special ceiling or sarking

The trussed rafter designer and/or supplier should provide his customer with certain information to enable the user to check the trussed rafter construction and provide him with information to assemble the roof structure on site. For further information please refer to section 11 (information required) in BS 5268: Part 3.

QUALITY CONTROL

Before proceeding with the construction of a trussed rafter roof, it would be helpful to understand the quality control imposed upon the manufacturer of trussed rafters. BS 5268: Part 3 section 8 'Fabrication', sets the standards for trussed rafter production, and this standard is incorporated in the Building Regulations 1991. All trussed rafters used for dwellings should be manufactured to this standard, whether fabricated using punched metal plate fasteners, nailable metal plates or plywood gusset joints. The illustrations in Figs. 5.3a–j attempt a graphical interpretation of part of section 6, but the reader is directed to the British Standard text itself for full information.

Figure 5.3a shows that moisture content of the timber used in fabrication should not exceed 22%.

Figure 5.3b covers the maximum gap allowed between two adjoining members under the punched metal plates. The average gap width should not exceed 1.5 mm unless specifically allowed for in the design.

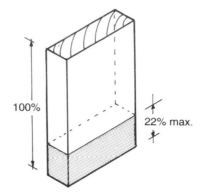

Fig. 5.3a Moisture content.

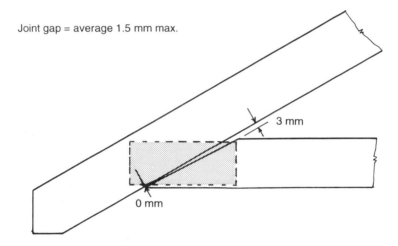

Fig. 5.3b Trussed rafter joint gap.

Figure 5.3c illustrates that wane, the term used to describe the occasionally occurring rounded corners of the timber caused by the timber being cut near to the outside of the log of the tree, is acceptable only at certain places on the trussed rafter. It is limited on the surfaces of the trussed rafter to which other elements of the building are attached, namely the top of the top chord and the underside of the bottom chord, and within the plate area of course no wane is tolerable.

Figure 5.3d shows that to ensure correct embedment of the plate teeth the difference in thickness between the members at a node point must not exceed 1 mm.

The permissible gap between the underside of the metal plate and the timber surface shown in Fig. 5.3e should not exceed 1 mm for nails or teeth up to 12 mm long, and 2 mm for nails or teeth over 12 mm long. This gap should not exceed 25% of the contact area of any member of the joint.

Figure 5.3f illustrates that timber is a living material and not man-made, and

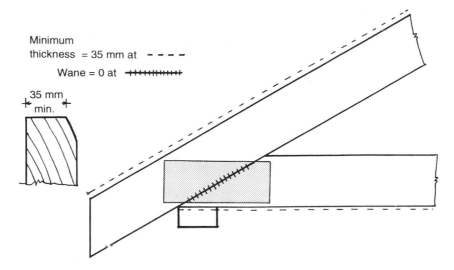

Fig. 5.3c Trussed rafter wane in joint area.

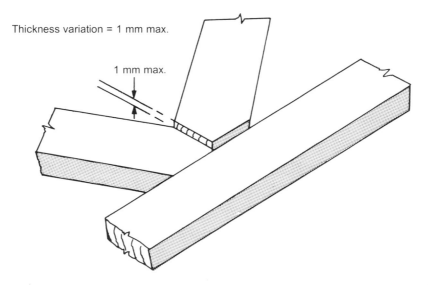

Fig. 5.3d Trussed rafter timber thickness at joint.

contains features which can detract from its overall strength. Such features are splits and fissures, live knots and dead knots and, in the latter case, possibly knot holes where the dead knot has fallen out of the timber section. Hairline fissures up to 55 mm long which have apparently been caused by the tooth or nail, are not seen as having any significant effect on the joint and can be ignored.

Figure 5.3g illustrates that corners of plates projecting beyond the edges of the trussed rafter timbers are not allowed. The metal plate or plywood gussets which project beyond the outer edges of the trussed rafter should have their protruding areas

Gaps not permitted (except close to thickness variation as above)

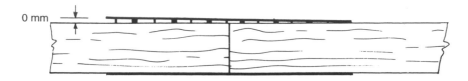

Fig. 5.3e Trussed rafter plate bedding.

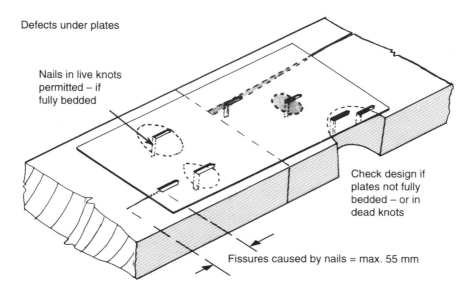

Fig. 5.3f Trussed rafter knots in plate area.

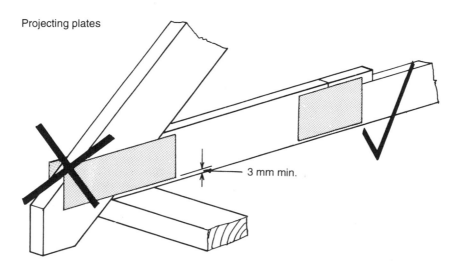

Fig. 5.3g Trussed rafter plate projections.

removed, or in the case of metal plates should be folded over once fully embedded. Particular attention should be paid to fasteners which protrude into walk spaces or other areas to which access may be gained at a later date. Where it is not possible to avoid a projecting edge, then a timber block must be placed between the plates to protect those in the roof space, and to protect those handling the components during construction.

Figure 5.3h illustrates plate location. Taking into account possible plate misplacement during manufacture (see laser location methods in Chapter 6), the British Standard in paragraph 8.2.1 states 'fastener misplacement during assembly should be within limits assumed in the design'. Unless a greater allowance has been made, fastener misplacement should be no more than 5 mm in any direction. These tolerances are essential and incorporate a safety factor in design.

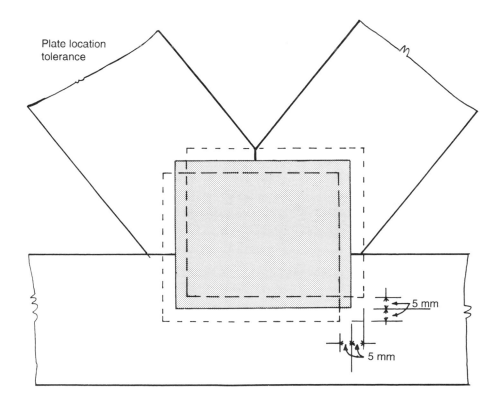

Fig. 5.3h Trussed rafter plate location tolerance.

Figure 5.3i illustrates dimensional tolerance. Most trussed rafters are manufactured in metal jigs, but because there is such a variety of trussed rafter spans, pitches and shapes, these jigs have to be quickly and easily adjustable. For this reason and bearing in mind the inherent natural movement of timber, it is essential that some tolerance from the design shape be allowed in production.

Figure 5.3j illustrates a marked trussed rafter. Until the publication of BS 5268: Part 3 it was not necessary to mark the trussed rafter with the name of the company

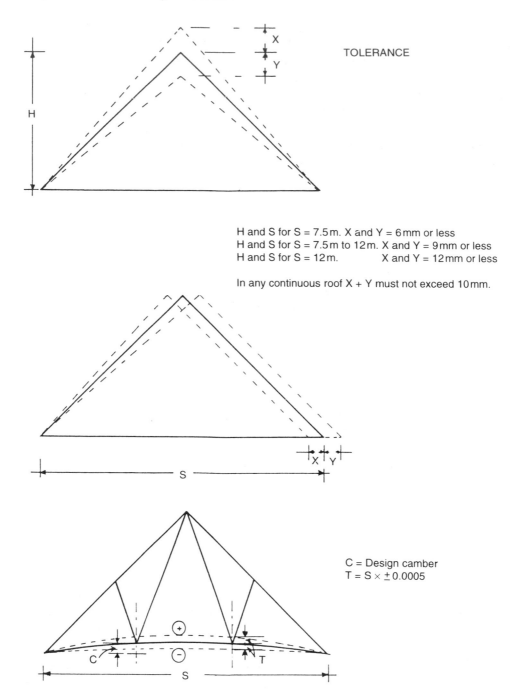

TOLERANCE

H and S for S = 7.5 m. X and Y = 6 mm or less
H and S for S = 7.5 m to 12 m. X and Y = 9 mm or less
H and S for S = 12 m.   X and Y = 12 mm or less

In any continuous roof X + Y must not exceed 10 mm.

C = Design camber
T = S × ± 0.0005

Fig. 5.3i Trussed rafter dimension tolerance. Where a camber is specified by a designer, it should be checked with the trussed rafter lying on its side before being fixed on to the roof structure. The measurement should be taken as the distance between a string line fixed between the wall plate support points where rafter meets bottom chord, and the underside of the nearest node point to mid span.

responsible for its manufacture. This has changed, with the requirement of the 1998 edition of the standard in section 8 'fabrication' in paragraph 8.1, under the heading 'marking', which requires that every trussed rafter should be clearly marked with the identification of the producer, the materials used and the standard to which it is produced. Furthermore the marking label or stamp should be placed as near to the apex as possible thus making it clearly visible in the completed roof void. Whilst it may be easier in manufacture to mark the truss near the wall plate support junction or on the bottom chord, this area is invariably covered with insulation and should anything go wrong at a later date with the roof structure then the vital information would be difficult to find.

Fig. 5.3j Trussed rafter marking.

INSPECTION

Clearly, as has been seen above, there are many aspects of trussed rafter production requiring a check to ensure their conformity with the British Standard. It must be said that the majority of trussed rafter producers are well aware of the standards required of them and produce trusses of high quality. To provide the specifier and the purchaser with an assured standard of quality, manufacturers must produce to the standards not only of BS 5268 but also those set down in ISO 9002, *Quality systems*. This ensures not only a conforming product but also minimum standards of service and administration that accompany the product. TRADA Certification Ltd operates such a scheme for trussed rafter manufacturers. Registered producers are allowed to use the BM TRADA certification logo on their product and documentation. The 'Q' mark indicates a quality assured product in the construction industry; this additional mark can only be used by

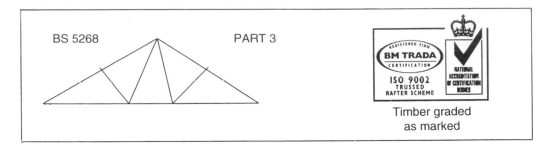

Fig. 5.3k Typical mark prior to 'Q' mark scheme.

firms who are registered to a 'Q' mark scheme. Figure 5.3k illustrates an earlier mark prior to the introduction of the 'Q' mark; this and similar marks will be found on trusses produced over the last 10–15 years.

Marks can be applied by ink stamping, as is often the case with the mark in Fig. 5.3k, or by fixing a tear resistant plastic label to each trussed rafter and trapping the label fixing area under the punched metal plates to ensure that it is not easily detached (Fig. 5.3j).

The BM TRADA scheme sets down standards incorporating those in BS 5268: Part 3 and ISO 9002. Each member has to record standards being achieved in respect of the criteria set down in BS 5268, logging the data as production proceeds. Random checks are carried out by BM TRADA inspectors at least twice each year to ensure that the standards are being met. Revisits are made more frequently if standards or records are not adequate, with the ability of BM TRADA to withdraw approval if the company fails to improve. Then the producer cannot prove that he is consistently producing trussed rafters to BS 5268. Effectively he can no longer produce trussed rafters which will prove acceptable to architects and building control supervised projects.

THE CONSTRUCTION OF A TRUSSED RAFTER ROOF

Guidelines to the correct handling and storage of timber and timber components are to be found in Chapter 7.

The erection of a simple trussed rafter roof is outlined in the TRA (Trussed Rafter Association) technical handbook. The method set out below follows those guidelines but expands the techniques and extends them to deal with intersecting roofs. BS 5268: Part 3: 1998 gives more detailed information on bracing domestic trussed rafter roofs. Care must be taken to check the criteria laid down in annex A of the British Standard are observed. The standard and amendment should be to hand when reading this section.

The roof is assembled generally with nailed joints. Galvanised nails must be used, the rough surface giving a better grip and the galvanising protecting against rust. All other steel work to be fitted into the roof will be galvanised, as are the punched metal plates on the trussed rafters.

A start should be made by studying the drawings and then by checking that all timbers, fixings and trussed rafters are to hand. Check also the spans, plates, pitch and overhang on the trussed rafter. Assuming that the wall plate is bedded and the mortar set, and that any plate straps have been fitted, the trussed rafter centres should be marked out on the plate. Reference should be made to the drawing for location of chimneys, loft access traps and other openings in the roof structure. On timber framed housing the precise setting out is usually given, in order to ensure that the trussed rafters are located directly over wall panel studs.

Check the span over the actual wall plates against the truss to be used. A note here on tolerance will be useful. An overspan of 20 mm, 10 mm each side, is a sensible tolerance on the trussed rafter, thus allowing some variation in line by the wall plate without making it necessary to notch the underside of the top chord of the truss to ensure correct bearing on the bottom chord. Larger errors may occur and limits of tolerance are set by British Standard 5268: Part 3: 1998. In Fig. 5.4, for pitches 35° or less, S2 must not exceed S1/3 or 50 mm, whichever is the greater. The new standard sets no limits for pitches above 35° and this in the author's opinion is an error and the author recommends the continued use of the limiting data set out in Fig. 5.4. On all pitches 'D' must not exceed B/2.

S1 = Length of scarf joint;
S2 = Bottom chord cantilever;
D = Width of wall plate not directly under connector plate;
B = Width of trussed rafter support

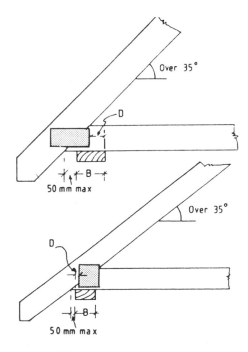
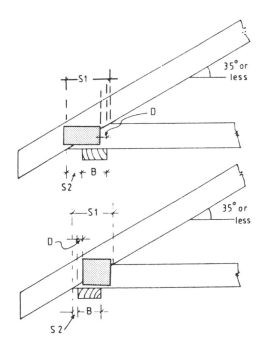

Fig. 5.4 Wall plate location tolerance.

Under no circumstances may a trussed rafter be cut or altered in any way on site without a design drawing and specification provided by the trussed rafter designer.

The practice of 'skew' or 'tosh' nailing the trussed rafter to the plate is not recommended, as this can easily damage both the heel joint timber and nail plates. Truss clips made from galvanised steel are available from most truss suppliers and builders' merchants and these should be fixed to the wall plate at the trussed rafter setting out marks. The truss clip is illustrated in Fig. 7.10. Under certain design conditions truss straps may also be required, but these could be fitted just prior to roof covering.

The object at this stage of construction is to erect a portion of the roof that will be stable, from which the remainder of the roof can be constructed. Diagonal braces should be used for stabilising the roof, *not the gable end* even if at this stage it has been built. The method now described will also work for the centre section of a hip roof. Reference should be made to Fig. 5.5.

The first trussed rafter, A, should be positioned a few trusses away from the gable or hip peak, such that its peak coincides with the top diagonal brace F. Holding this truss truly vertical, temporary diagonal brace B should be fitted on both sides of the roof. The brace should be well nailed to the trussed rafter as close to the web joint as possible but not in it, and to the wall plate. Temporary fixings should be made using double-headed shuttering nails to avoid damaging timbers when they are moved.

Prepare a temporary batten for both sides of the roof, long enough to reach from trussed rafter A to the gable. The battens should be marked with the truss centres by reference to the wall plate marks and shuttering nails driven in until the points just show through. Before standing truss C in place, ensure that it is the correct way round to match truss A. This can be done by checking that the bottom chord splice plates will be in-line down the length of the roof. If no splice plate is fitted, then check the truss for similarity whilst still in stack. *Do not hand the trusses*; failure to carry out this check could result in poor roof alignment caused by the peak being slightly off true centre. Stand trussed rafter C in position and fix to wall plate with truss clip. Fit the temporary battens to trussed rafters A and C on both sides of the roof; the prefixed nails now easily locate the trusses and provide a free hand to help stabilise the structure. The remainder of the trussed rafters in this first section can now be fixed, each nailed to the temporary batten on both sides of the roof.

Before proceeding further, check that the trussed rafters are truly vertical. Acceptable tolerances for plumb are set out in BS 5268: Part 3: 1998, section 9, paragraph 9.3.1.

MAXIMUM DEVIATION FROM VERTICAL

Rise of trussed rafters (m)	1	2	3	4	or more
Deviation from vertical (mm)	10	15	20	25	

Before the current British Standard there was no specification for roof bracing timbers, except the overall section size. The latest British Standard in annex A(g) sets down a minimum width of 89 mm and a minimum depth of 22 mm. The timber must be of a species listed in Table 1 of the standard, and free from major strength reducing

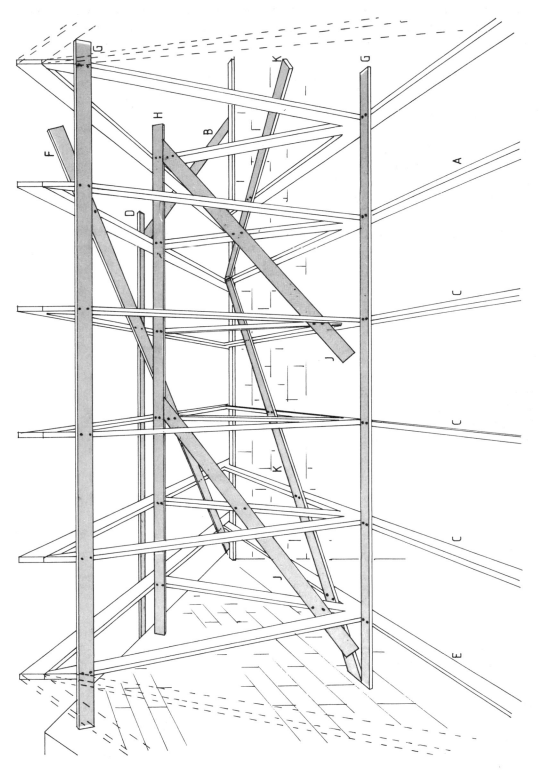

Fig. 5.5 Trussed rafter stability bracing.

defects. The cross section should not be less than 2134 mm². This means minimum bracing sizes can be *22* mm × 97 mm or 24 mm × *89* mm. No other strength grading is required.

Annex A(h) of the standard also stipulates that the bracing members must be nailed to the trussed rafter with 2 × 3.35 mm diameter galvanised wire nails on every crossing point. The minimum nail length should be equal to the bracing thickness plus 32 mm, i.e. 54 mm for 22 mm thick bracing or 56 mm for 24 mm thick bracing. In practice the nearest readily available galvanised nail of 3.35 mm diameter is 65 mm long and nail specification has been general practice for many years. For the purposes of the following description we will use 22 mm × 97 mm bracing and 3.35 mm diameter galvanised wire nails 65 mm long.

Diagonal brace F, 22 mm × 97 mm, should now be fitted to the underside of the top chords of the trusses on both sides of the roof at an angle of approximately 45° to the rafters. At least four diagonals 'F' should be fitted to each roof. Two 3.35 mm diameter × 65 mm long galvanised round wire nails should be used at each trussed rafter where the diagonals cross. The diagonals must extend down the wall plate, and should be let in to the top of the plate and nailed as above. If the brace has to be lapped in its length, then the lap should cover two trussed rafters and each piece be nailed to the trussed rafters as described. Figure 5.6 illustrates the diagonal brace lap, viewed from beneath the rafters; other binders are omitted for clarity.

Longitudinal binders G, again 22 mm × 97 mm in section, should now be fitted at peak and third point. Nailing is the same as for the diagonal, but check the centres of the trusses and the bottom chord alignment before fixing. Failure to do this could result in ceiling fixing problems – see Fig. 5.7, the trusses illustrated, not being located at precisely 600 mm or 400 mm centres, will not readily accept ceiling boards which are generally 1200 mm wide. This necessitates a batten being nailed to the side of the trussed rafters to correct the spacing error. The ends of these binders should be tight against the gable wall. On timber framed housing binders should be nailed to the gable end panel. Blocking pieces will be required, the depth of the bottom chord sufficient to enable the third point binders to be fixed to the panel frame.

On most domestic roofs the binders and bracing described thus far will be adequate to stabilise temporarily the trussed rafters themselves, at one end of the roof. The procedure is repeated at the opposite end of the roof, creating two stable elements, and finally the centre section is filled in, taking care to lap the longitudinal G over two trussed rafters.

The roof structure has to absorb the load from wind blowing on the gable ends, causing pressure at one end and suction at the other. BS 5268: Part 3, appendix A gives standard bracing, provided certain conditions of wind speed, pitch and span are not exceeded. Specification must therefore be checked before proceeding.

Continuing with the construction, it will be assumed that the standard wind bracing is satisfactory. At least four diagonal bracers F must be fitted at approximately 45° to form 'X' formations down the whole length of the building. The longitudinal binder H should be fitted, timber section and nailing as before, from one gable to the other, lapping as required. Fit this to both slopes of the roof close to the web on top chord

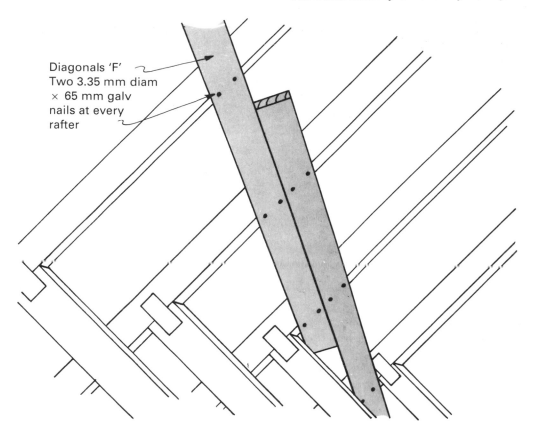

Diagonals 'F'
Two 3.35 mm diam
× 65 mm galv
nails at every
rafter

Fig. 5.6 Diagonal bracing laps.

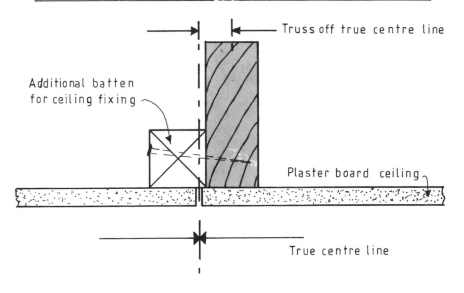

Truss off true centre line

Additional batten
for ceiling fixing

Plaster board ceiling

True centre line

Fig. 5.7 Trussed rafter location error.

joints. Diagonals J should now be fitted to the short webs, but only on trusses greater than 8 m span. Although not critical, the direction of slope should be from the bottom chord at the gable to the top chord; the slope will reverse at the centre line of the building. Finally, diagonal K should be fitted on top of the bottom chords; these should be placed at approximately 45° to the bottom chords and produce a 'W' formation from one gable to the other. Diagonal 'K' is not a requirement of BS 5268: Part 3: 1998, but in the author's opinion is advisable. Annex A of the British Standard recognises the lack of lateral stiffness of house supporting walls exceeding 9 m long between buttressing, and restricts its bracing information accordingly. Many houses at construction stage have walls on which trusses are supported with no buttressing from internal partitions, and are no more than a series of slender columns of cavity brickwork connected only by steel or concrete lintels. At roof construction and tiling stage the roof does not have the benefit of the plasterboard ceiling diaphragm. In timber framed housing brace K is essential.

The roof is now capable of withstanding the wind loads placed upon it, but the gable walls are not connected to it. On the pressure side the wall would be blown onto the roof and would probably be secure, provided the blocking had been fixed between the wall and the first truss, and between the first and second trusses (see Fig. 7.18). On the suction side, the wall would be sucked out. The gable walls must therefore be fixed securely into the roof structure with metal straps. These gable restraint straps apply to all forms of roof construction and are therefore detailed in Chapter 7 under a separate heading. British Standard 5268: Part 3, appendix B gives an appropriate standard for trussed roofs. Details of strapping roofs to walls are also given in Approved Document A of the 1991 building regulations, in the NHBC chapter 7.2 'pitched roofs' and in the Trussed Rafter Association technical leaflet.

To complete the roof structure, gable ladders, water tank platform, barge, fascia, soffit, ventilators and eaves insulation controllers must be fitted. These items are dealt with in detail in Chapter 7. In addition to the bracing A–K described above, some extra bracing may be required by the truss designer, to prevent compression buckling of long web members. This additional bracing will take the form of more rows of brace H, but fitted at the centre of the web length. These bracing rows also require a locking brace (as J), fitted to the opposite face of the web at approximately 6.0 m centres.

HIP END ROOFS

The construction of trussed rafter hip end roofs varies from one truss system to another. Precise construction details are therefore given in Chapter 6 under each plate manufacturer's heading.

On roofs less than 5 m span almost traditional hip construction may be used, as detailed in Chapter 3. In excess of this span most systems use prefabricated hip trusses with, in some cases, only the extreme lower corners of the hip being of site cut and fitted construction. More details of the variation between trussed rafter manufacturers' hip construction are given in Chapter 6.

Rafter diagonal bracing is not generally required in a hip roof, the hip itself being a rigid diagonal structure. Longitudinal binders G must be continued from the main roof into the hip construction, plus additional binders H fixed flat on top of the hip trusses' top chords. Diagonal K should be installed (see Fig. 5.5).

Most trussed rafter hip constructions involve a compound girder assembled from two, three or in some cases four individual trussed rafters. The nailing together of these individual components to form a girder is critical, and should be carried out by the trussed rafter manufacturer at works. If girders are not nailed by the manufacturer then a nailing and bolting detail must be provided by the roof designer. When fastened together on site, bolts must be used for the bottom chord members, in positions marked by the truss manufacturer. In all cases the nails and bolts must be positioned strictly in accordance with the manufacturer's instructions. See a typical nailing and bolting pattern given in Fig. 5.10a. The setting out of the girder and intermediate trussed rafters is also critical and dimensions given on drawings must be carefully followed.

VALLEYS AND INTERSECTING ROOFS

The construction of valleys on trussed rafter roofs varies little between plate systems, most using a set of reducing trusses imposed upon the main roof structure. The designs are however dealt with separately under each system heading in Chapter 6.

Complex intersections normally have special design drawings supplied for them by the trussed rafter manufacturer. The solutions follow a pattern and typical details are given below. Firstly, Fig. 5.8 illustrates the construction of a roof intersection on which the main roof cannot be supported on a wall or beam, but is instead supported on a girder or compound truss. Secondly, in Fig. 5.11, a further complication is caused by a hip occurring at the intersection.

The location of the girder truss must be such that it will carry the main roof standard truss, the overhang of the standard trussed rafters simply being removed on site if special trusses have not been provided. If the overhang is removed on site, care must be taken not to cut too near the heel joint plate: a minimum distance of 10 mm is recommended. See Fig. 5.9 for the girder to standard truss detail.

Girder or compound trusses are made up of two or more specifically designed trussed rafters. The girder has to carry considerable loads from the end of the standard trussed rafters of the main roof, the load concentrated on the bottom chord. To stiffen this member, a deeper timber section is often used and a different configuration of webs dividing the bottom chord into four or more bays, such as those shown in Fig. 2.8. The girder should be fully assembled in the factory, arriving on site as one unit not separate trusses. If this is not the case, then a nailing and bolting specification must be supplied by the trussed rafter manufacturer. The correct nails and bolts and spacing must be used if the individual trussed rafters in the girder are to function as one structural unit. The load from the main roof is concentrated on one side of the girder and inadequate fixing will lead to deflection giving a poor ceiling line and, more seriously, possibly a failure (see Fig. 5.10b).

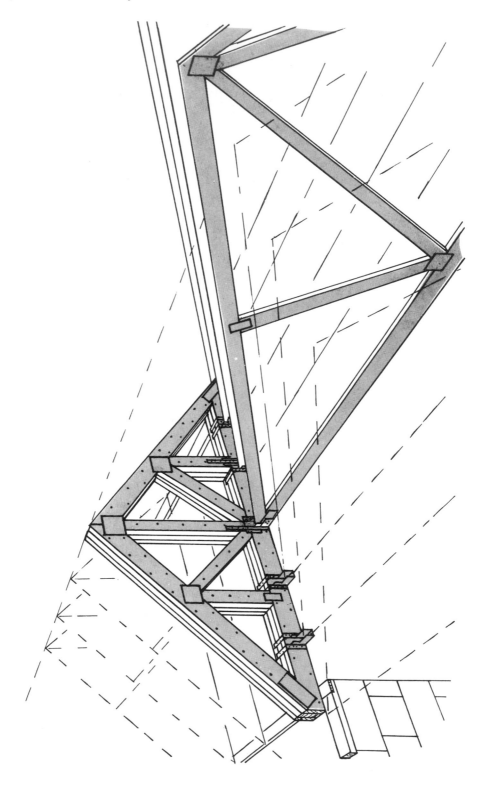

Fig. 5.8 Trussed rafter valley girder.

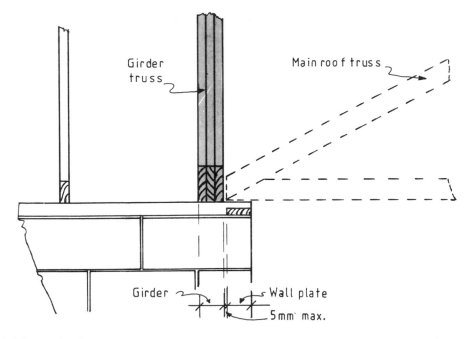

Fig. 5.9 Trussed rafter valley girder location.

The shoes used to support the main roof trussed rafters on the girder must be designed for the function. Normal joist hangers are not adequate for two reasons:

(1) They are unlikely to be strong enough.
(2) They will have insufficient bearing for the truss. If a 100 mm wide wall plate is needed, then 100 mm bearing is needed in the shoe. As a guide the bearing should

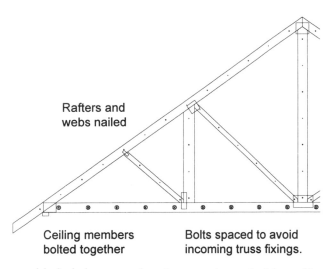

Fig. 5.10a Site assembled girder truss showing mandatory bolting of bottom chord.

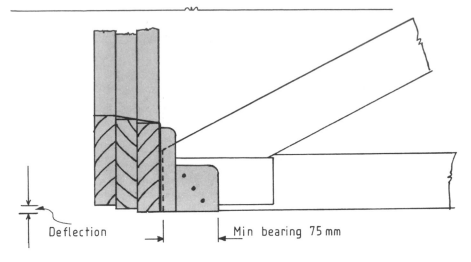

Fig. 5.10b Defective girder fixing.

not be less than 0.008 times the span of the trussed rafter, with a minimum value of 75 mm, unless of course structural design can show a smaller bearing is adequate to avoid crushing the underside of the trussed rafter bottom chord, or the wall plate.

Standard truss shoes are available from most trussed rafter manufacturers and builders' merchants. The shoes must be fixed to the girder with the correct number and size of nails, screws or bolts specified.

Referring now to the hip end valley intersection illustrated in Fig. 5.11, the loading on the girder is even further complicated by the large point load from the hip girder truss. A special galvanised steel shoe is required to transfer this load from the hip girder to the supporting girder. Fixing must be by bolts, screws and nails being inadequate. Bolts should not be placed in the area of the connector plates, one solution being to bolt above the junction through one of the girder webs (see Fig. 5.12).

TRIMMINGS FOR OPENINGS

Small openings for flues or loft hatches are dealt with in Chapter 7 under the appropriate headings. Details below apply to large openings, i.e. those greater than two standard trussed rafter spacings. Such openings can be provided using specially designed girder or compound trusses either side of the opening, or by taking support from the structure itself, in which case special trusses will be required for the infill area.

On the true trimmed opening, a purlin or purlins must be fitted between the girders onto which common rafters are fixed. The ceiling joists will be supported on a purlin at bottom chord level on the girder truss. All purlins and binders must be fixed to the trussed rafters at the node or joint points. It is good practice to have the common infill

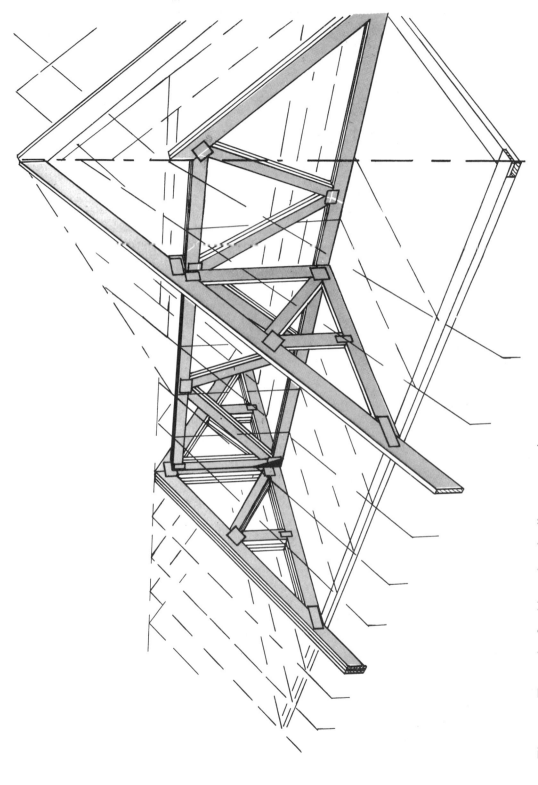

Fig. 5.11 Trussed rafter hip end and valley construction.

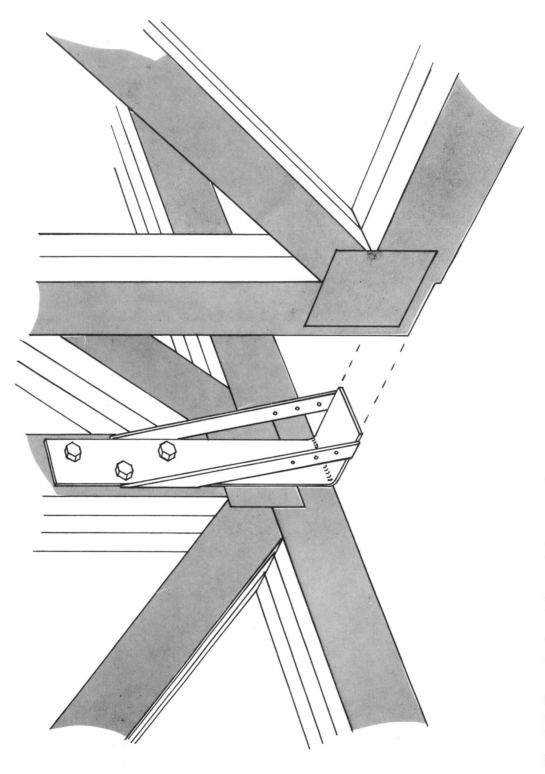

Fig. 5.12 Trussed rafter girder to girder connection.

rafters 25 mm deeper than the top chord of the truss to allow for birdsmouthing over wall plate and purlin (see Fig. 5.13).

The alternative method, using the structure passing through the roof for support, is suitable for particularly large openings but does presuppose that the structure is built in advance. The special shaped trusses are shown in Fig. 5.14 and can be supported either on steel shoes built into the structure or on a corbel. Care must be taken to stabilise the upper part of the special truss; this should be done with both binders and diagonal bracing.

Note that timber located adjacent to flues must conform to the rules laid down in the Building Regulations concerning the proximity of combustible materials. See the section in Chapter 7 of this book, 'Trimming small openings'.

ATTIC AND LOFT ROOFS

For the purpose of this section an attic will be defined as a roof void used for living, bedroom, bathroom and kitchen, playroom or studio use, whilst loft will be a roof void intended for storage only.

One of the disadvantages of the normal fink or fan trussed rafter roof is that the many web members obstruct easy access within the roof void, a space traditionally used for storage of household items. The traditional purlin and common rafter roof described in Chapter 3 gives a relatively unobstructed loft space, although the collars may be well below a desirable headroom height. The TRADA roof described in Chapter 4 also gives a very usable loft void.

The trussed rafter roof can provide either a loft or an attic, if specifically designed to do so. If the roof is designed with a loft space which the designer knows, or because of the shape of the internal members of the truss, can provide a room, then he should allow for domestic loading.

We now consider four main types of loft or attic roof:

(1) Loft void – trussed rafters carrying roof load with separate joists for storage load.
(2) Loft void – trussed rafters carrying roof and floor load.
(3) Attic – trussed rafters carrying most of roof load with separate joists for floor loads.
(4) Attic – trussed rafters carrying floor and roof loads.

(1) For special trusses designed to provide an unobstructed roof void for some part of the roof, the truss loading is as for standard trusses. In between the trusses, independent joists are installed, taking support on the same wall plate as the trussed rafter and on some intermediate load bearing wall (see Fig. 5.15). The timber sections used in the trussed rafters will be standard and not the heavier sections required with the full attic, thus the cost of the truss is not greatly increased. Care must be taken to add the necessary herringbone strutting, and the strutting indicated in the illustration on the intermediate wall plate to brace the infilling floor joists. This particular construction is

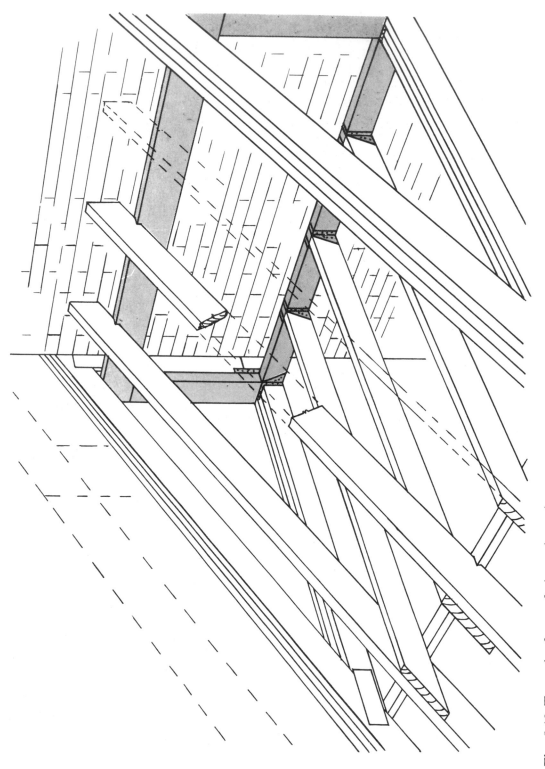

Fig. 5.13 Trussed rafter roof trimmed opening.

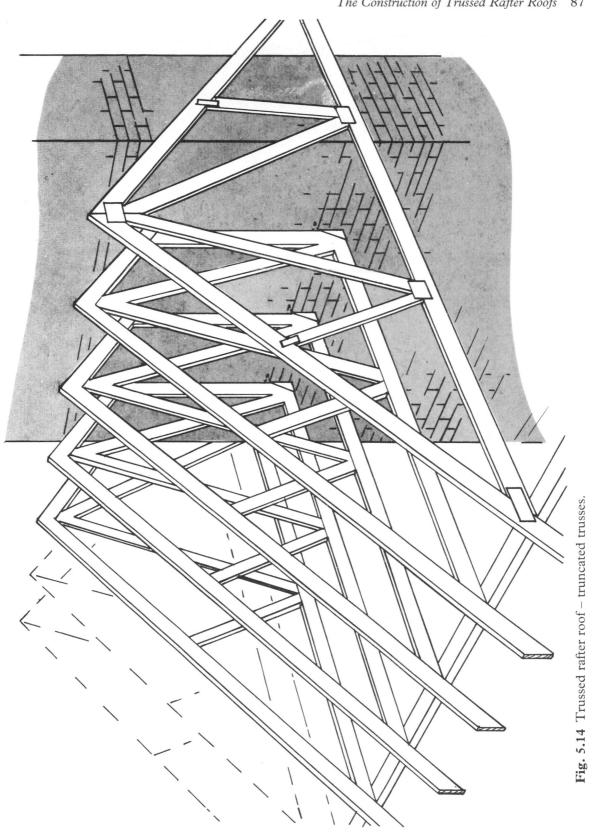

Fig. 5.14 Trussed rafter roof – truncated trusses.

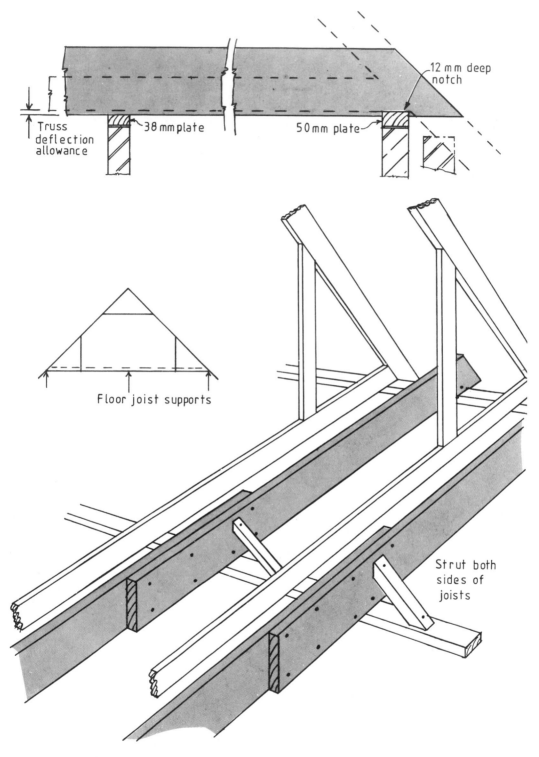

Fig. 5.15 Loft floor support.

recommended where an existing trussed rafter roof is being used to carry more than simple loft storage loading.

(2) A specially designed trussed rafter, designed to carry both roof and floor loads, with or without internal support for the floor. To avoid the truss becoming almost an attic, thus containing costs, the loaded area of the loft should be kept as small as is practical, particularly on the clear span design. In both cases, the trussed rafter designer may seek to close the trussed rafter spacing down to 450 mm or 400 mm centres. Figure 5.16 illustrates this loft option. If insulation is to be fitted between the rafters, consideration must be given to ventilation (see Chapter 7).

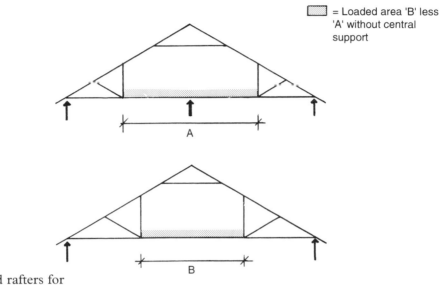

= Loaded area 'B' less 'A' without central support

Fig. 5.16 Trussed rafters for loft loading.

(3) The attic design solution, given below, is particularly suited to dwellings where the span is large but the distance between gable walls is relatively narrow. Choosing the shortest span for a roof structure usually results in greatest economy. This design uses the cross-walls to support purlins onto which trussed rafters and floor joists are fixed. The design also allows the use of wide dormer windows without the need for heavier trimmers, and simplifies forming openings for stairwells. It is, however, labour intensive compared to option (4). Figure 5.17 illustrates option (3).

(4) Figure 5.18 shows a fully independent attic trussed rafter construction. The timbers used will now be much larger in section than for standard rafters, thus increasing the unit weight; this must be considered for handling reasons. The steeper pitch necessary to allow satisfactory room within the roof can mean that the height of the truss will be outside both manufacturing and transportation practical limits, and in such instances the roof truss may be split horizontally into two components, the split occurring at the ceiling joist or collar position. Producing two trussed rafters to form each frame therefore adds to the cost of such construction. The overall economy of the

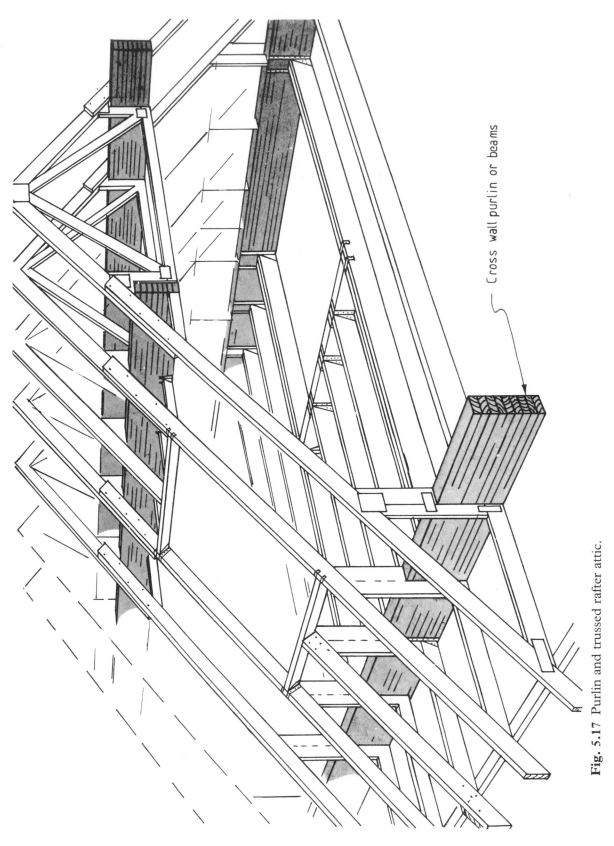

Cross wall purlin or beams

Fig. 5.17 Purlin and trussed rafter attic.

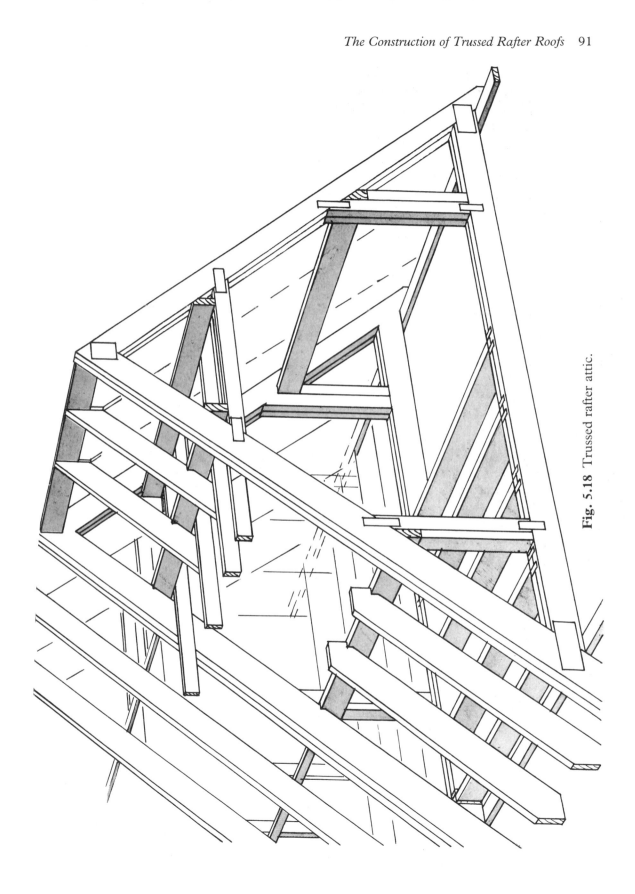

Fig. 5.18 Trussed rafter attic.

attic trussed rafter roof is very dependent on dormer or roof window and stairwell planning. This aspect is now discussed.

If a clear spanning attic trussed rafter is to be used then there is complete freedom of roof planning, both at attic level and on the floor below. However, location of stairwell and dormer or roof windows can have a dramatic effect on the cost of the attic structure. Each trimmed opening will require girder trussed rafters; the wider the opening the more trussed rafters to each girder. Bearing in mind that each attic trussed rafter will cost approximately four times its standard non-attic equivalent, it can be seen that girders must be kept to a minimum. Set out below are five basic rules of attic trussed rafter roof design economy.

(1) Plan to keep trussed rafters on a 600 mm spacing.
(2) Plan the stairwell to run parallel with the bottom chord and keep as narrow as possible.
(3) Plan dormer and roof windows as narrow as practical (some roof windows will fit between 600 mm centre trusses).
(4) Plan dormers and windows in-line across the building.
(5) Attempt to keep the overall height of the truss within transportable and/or manufacturing limits. This is often around 4 m, avoiding the additional cost of having to split the truss into two sections. Figure 5.19 illustrates a split truss.

Figure 5.20a illustrates a poor attic layout involving many heavy girders, few trussed rafters at 600 mm centres, and much site fixed infill and should be avoided. Figure 5.20b, however, is a much more economical layout involving lighter girders, more trusses at 600 mm centres and less site infill. The layout in Fig. 5.20a could be solved using the trussed rafter, purlin and joist design shown in Fig. 5.17. Intermediate support may be required for the purlins, with the floor joists being supported by further purlins or directly off walls below.

OPENINGS FOR DORMERS, ROOF WINDOWS AND STAIRWELLS IN ATTIC TRUSSED RAFTER ROOFS

Openings in attic trussed rafter roofs can be formed in a similar manner to those shown in the traditionally constructed roof section, Chapter 3. Using two or more trussed rafters as trimmers, both infilled rafters and floor joists need to be supported from those trimming trusses. Short purlins are used to pick up the ends of the trimmed rafters, the purlins themselves being supported on girder trusses. The purlins should be supported on purlin posts built in to the attic girders at the lower purlin level, and supported on the ceiling tiles as close to the joint between top chord and ceiling tie as possible for the upper purlin. The purlin should be notched under the top chords to give an adequate birdsmouth for the oncoming infill rafters. The purlins will of course carry the infill rafters, leaving the floor joists to be infilled. This can be done using short lengths of floor joists of a matching depth to the bottom chord of the trussed rafter fixed across the

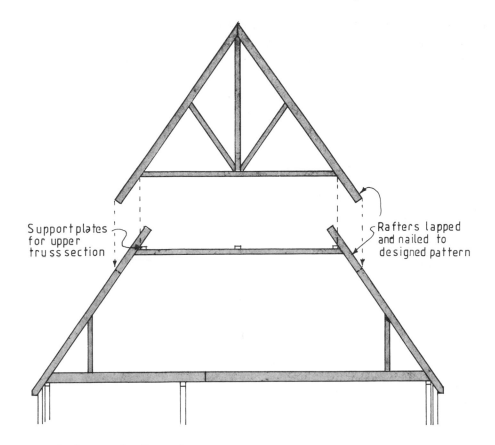

Fig. 5.19 'Top hat' trussed rafter attic.

opening, the infill joists being supported in joist hangers fixed to the sides of the bottom chord.

Stairwell openings can be similarly constructed, but in this case the whole of both rafter slopes will need supporting on purlins as described above. Figures 5.17 and 5.18 illustrate typical trimmed openings in two types of trussed attic construction. If the sloping ceiling in the infill area between the girder trusses need not align with the main roof, then the purlin can be lifted, and lighter smaller sectioned infill rafters used. On narrow openings between girder trusses it is practical to use joist hangers as purlin supports, fixing these directly to the sides of the top chords of the girder trusses.

As can be seen in Figs 5.20a and 5.20b, the positioning and size of openings can significantly affect the cost of constructing trussed rafter attic roofs. Attention is therefore drawn to the design in Fig. 5.17, which, because it is based on large purlins, allows great freedom of creating openings of any shape or size between the upper and lower purlin without structural complications. If a large number of openings, or alternatively extremely large individual openings are required, then this option is strongly recommended.

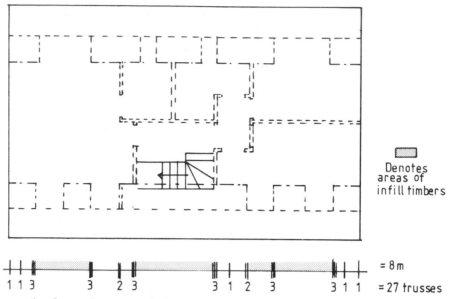

Denotes
areas of
infill timbers

= 8 m

1 1 3   3   2 3   3 1 2 3   3 1 1   = 27 trusses

Fig. 5.20a Trussed rafter attic – poor design.

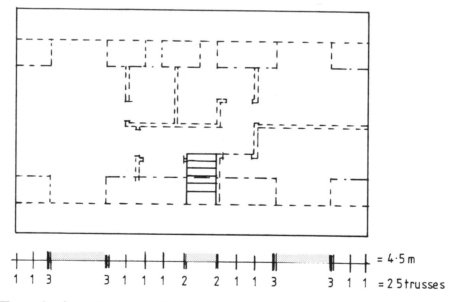

= 4·5 m

1 1 3   3 1 1 1 2   2 1 1 3   3 1 1   = 25 trusses

Fig. 5.20b Trussed rafter attic – good design.

BRACING THE ATTIC TRUSSED RAFTER ROOF

The trussed rafter attic roof must be constructed strictly in accordance with the designer's and manufacturer's drawings. It must be fully appreciated that the structure about to be constructed forms almost 50% of the new home and will carry far more load than the conventional roof structure. The trussed rafters impart great rigidity across the building within themselves, but rely purely on the bracing construction between them for their lateral stability, and therefore the lateral stability of the whole building above the wall plate position. It is *not* correct to consider that brick and block gable ends will impart any support for the roof structure in this lateral direction, indeed the timber roof structure itself must be strong enough to restrain adequately the large gable end areas.

A preliminary consideration of bracing is given below, but the more detailed implications of correct bracing are dealt with in Chapter 7 where they cover both traditional and trussed rafter attic roofs.

Temporary bracing of these larger heavier trussed rafter components is vital. The temporary diagonal brace placed on the top of the top chord should be at least 22 mm × 97 mm in section and well nailed to both plate and the trussed rafters as far up the rafter as practical. On a single-storey building it may be possible to place a temporary support, or prop, from the floor slab to the vertical side wall and top chord junction of the attic trussed rafter for additional security. Erection should proceed basically as for the standard trussed rafter roof, except that temporary diagonal bracing should be added to the underside of the top chord until permanent bracing of the roof has been achieved. This temporary bracing may well have to be fixed on the underside of the top chord or rafter within what will become the room area.

Longitudinal binders should be fitted generally as before, but there will be more of them and these are indicated in Fig. 5.21. This illustration does *not* show all of the bracing required in an attic roof; further reference should be made to Chapter 7, Fig. 7.22. The floor boarding will of course eventually act as a substantial binder for the bottom chord, therefore only temporary binders need to be installed prior to the laying of the floor itself. These binders are essential if correct spacing of the trusses is to be

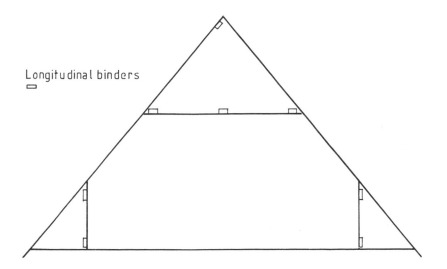

Fig. 5.21 Trussed rafter attic bracing. See Chapter 7 'Bracing' for full bracing details.

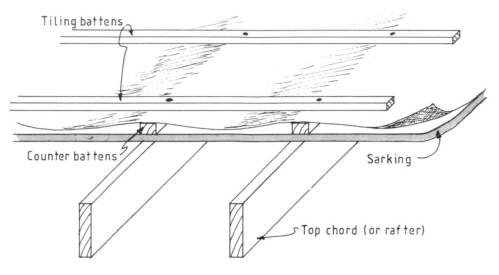

Fig. 5.22 Roof sarking.

maintained, which will of course assist cutting and fitting of the floor boards at a later date.

The problem of the lateral stability of the room remains to be solved on a permanent basis. In Scotland, Scandinavia and the USA where roof sarking is common, this forms a substantial brace and nothing further is required. The sarking often takes the form of a sheet material nailed to the top of the rafters over the whole roof area (see Fig. 5.22). BS 5268: Part 3: 1998, annex A, section A2, gives guidance on suitable sarking materials and fixing.

For information on loft to attic conversion work see Chapters 8 to 11 inclusive.

TRUSSED RAFTER SHAPES

Other common trussed rafter shapes are shown in Fig. 5.23.

- A Fink – equal pitch;
- B Stubb fink or bob-tail;
- C Cantilever fink;
- D Northlight;
- E Horizontal split truss;
- F Monopitch;
- G Asymmetric – unequal pitch;
- H Inverted fink;
- I   Raised tie;
- J   Horizontal split attic.

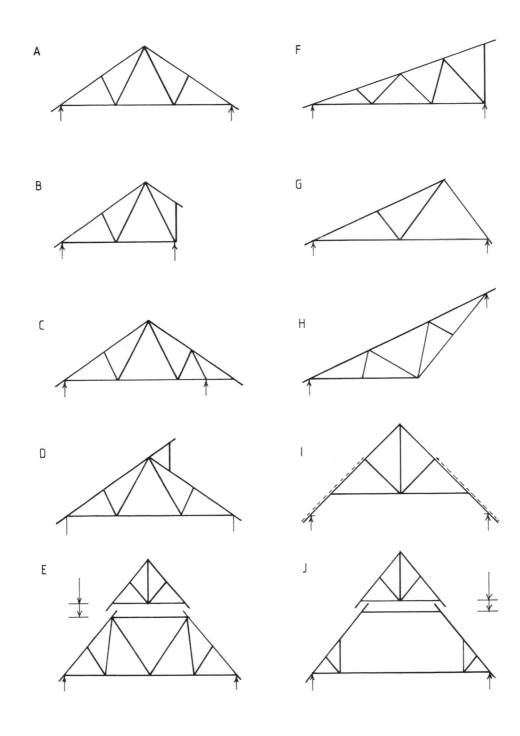

Fig. 5.23 Typical trussed rafter shapes.

CHAPTER 6
Truss Plate Systems

Most users of trussed rafters will first come into contact with the name of one of the plate system manufacturers on receipt of a set of computer printed calculations. If these are not asked for, it is possible that the user may never know which brand of plate has been used in the manufacture of the trussed rafter supplied to him.

The plate manufacturers or 'system owners' supply both plates and design information to trussed rafter manufacturers who, clearly, are free to choose which system they prefer. The choice will be made on plate price, but more importantly (and increasingly a deciding factor) on the degree of design information and technical back-up available from the system owner.

Whilst at first sight the type of plate used on the trussed rafter may seem of little importance to the end user, because his contract to purchase trussed rafters is with the trussed rafter manufacturer, it can affect both the quality of service and product supplied – and more importantly the precise construction form used on the roof.

The purchaser should be interested because the system owners, through their various manufacturers, can offer different forms of roof construction, some of which may appear very competitive on first sight but may involve a significantly higher site labour input than an alternative offer. The quality of the fabrication of the trussed rafter is another factor and this, to some extent, will be affected by the machinery and cutting data made available by the system owner to the fabricator. The purchaser is advised to ensure that his supplier is a member of the BM TRADA certification scheme and that the trusses are marked accordingly (see Fig. 5.3j and 5.3k).

SYSTEMS AVAILABLE

Changes involving the producers of the punched metal plates or system owners continue. Gang-Nail, the original in the UK, is still producing. MiTek Industries Ltd, a combination of Bevplate and Hydro-Air, is still a major producer but has disposed of its builder's fixings to Simpson Strong-tie. Wolf Systems Ltd was a newcomer when the last edition of this book was published, but since then has gone from strength to strength. In the author's opinion this has been based on its superior 'user friendly' computer engineering, not only for the trussed rafters themselves but for the whole roof design software.

The most recent change is the take-over of Truss Wall/Twinaplate by Alpine. Like most of the plate producers, Alpine is an international company and like all of the others has decades of experience to draw on. Although some names of plate producers have changed, a core of experienced people remain within the industry; however, some have moved with the mergers and take-overs which have occurred. The author has known and worked with some of the senior engineers who are now running these companies for over 30 years.

Wolf, MiTek, and Gang-Nail all now have web sites available for the provision of information off the 'Net', with MiTek's probably being the most developed at present, although undoubtedly all of the plate producers will have comprehensive web sites in the near future. Web addresses of those which currently exist as this book goes to press are set out in the bibliography section.

The amalgamation of some of the system owners is the result of the need to pool resources, particularly as the emphasis in system owners' success has continued to shift towards the need to provide increasingly more sophisticated computer programs for their trussed rafter fabricators to use. The cost of producing these programs can only be recovered by the sale of the nail plates. The competition for plate customers – the trussed rafter manufacturers themselves – is therefore fierce, continually driving the system owners to be 'the best' and thus attract more customers.

Most people know the name Gang-Nail, this being the generic term for punched metal plate connectors and indeed the company has the longest history of plate production in this country. Each plate producer has to conform to the standards set down in BS 5268: Part 3: 1998 and have a current certificate of conformity produced by a government approved testing and assessment authority for each type of plate produced by them. Historically this has meant an Agrément Board certificate, but now other private assessments bodies are authorised for this work and MiTek Industries has chosen Wimlas Ltd as their approving body. Other truss plate producers seem to have remained with the Agrément Board.

Gang-Nail produces both 14 and 20 gauge plates, with Wolf and MiTek producing 18 and 20 gauge plates. Alpine have a 20 and 16 gauge plate. All plates have to be marked with the manufacturer's name. Immediately after take-over, as is now the case with Alpine, there may be some of the incorporated company's plates still in circulation.

All companies have stainless steel plates available for trussed rafters for use in aggressive environments such as agriculture and some industrial situations. Standard plates are made from galvanised mild steel strip, this being simply punched with each producer's individual nail-tooth pattern and shape and cut to size. They are generally coated with a protective oil (which also aids production), boxed and delivered to the trussed rafter manufacturer.

Most trussed rafter plate producers mark their plates by stamping the system name –

some with colour for ease of identification. A wide range of plate sizes is produced by all manufacturers to cope with varying timber sizes and load stress.

The nail shape and pattern vary from one producer to another and affect the strength the plate can generate in the joint. Development of the nail profile, pattern and steel specification can therefore affect the size and cost of the plate needed to perform the given structural specification. This in turn will affect the cost of the plates on a truss and therefore the plate supplier's customers' competitiveness in seeking trussed rafter business. Gang-Nail, for instance, has its GN80X plate available; this uses a special high-strength steel in 20 gauge and is particularly useful in reducing the cost of spliced joints. These are the end-to-end butt joints necessary in some top and bottom chords to give an adequate length of timber.

COMPUTER PROGRAMS

The trussed rafter industry is one of the most advanced producers of building components in the UK in terms of its use of computer-aided design, drawing and production information. All plate producers supply their manufacturers with sophisticated software for use on personal computers, the PCs having become increasingly powerful in terms of their ability to store large quantities of data and process this at high speed. Programs now provide:

(1) The structural calculation of endless configurations of trussed rafters to easily establish the most economic solution with simple editing by the user to examine the 'what if . . .?' alternatives.
(2) Production of quotation from the selected design.
(3) Production of that quotation in hard copy form by word processing combined within the software for direct issue to the customer.
(4) Generation of full engineering calculations for customers to issue to their clients for Building Regulation approval.
(5) Computer-aided drafting produced from simplified input, to show roof layouts and roof types and connection details. Many of these standard constructional points are drawn from a library in the computer data bank. Figure 6.1 illustrates a typical part layout from Wolf's drafting software. This can be produced in black and white or colour, the latter of course helping to make the drawing easier to understand.

In addition, all now have general design packages to assist the roof designer with calculation and a selection of appropriate additional support members such as purlins and heavy beams. Alpine has a program they call *Tool-box* which includes the above and allows for ply-gusseted joints in the design as well as other engineering aids.
(6) Manufacturing data including:
(a) order picking, number of length and size and grade of timber, number and size and specification of plates;

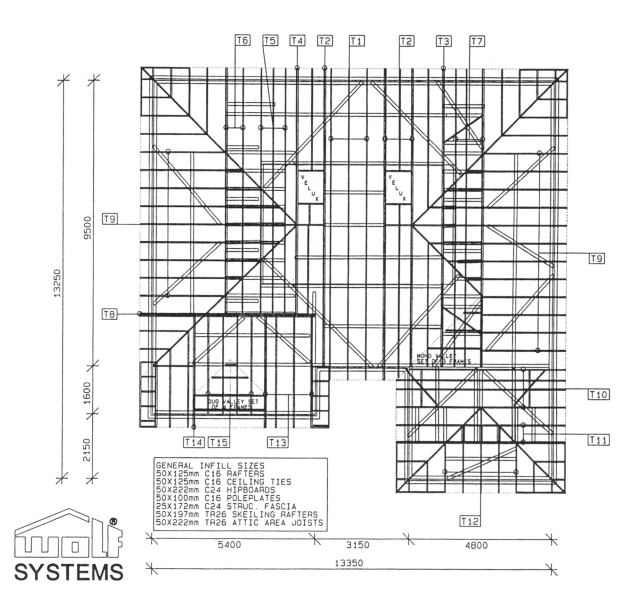

Fig. 6.1 Computer-aided drafting roof layout.

(b) precise cutting data for each member of the trussed rafter including length to the nearest millimetre and all angles and grade of timber (see Fig. 6.2);

(c) plating data on each joint indicating plate size and its precise location on the joint;

(d) jig setting;

(e) factory management data relating to production scheduling; some can go on to stock control, giving automatic re-ordering levels for timber and plates.

E-mail is now used by many plate producers and the larger manufacturers for instant data transmission. The computer generated cutting data for instance can be fed directly from some centralised design office to the CNC component saws in the factories of satellite manufacturers. This creates an almost paperless production process and of course eradicates the errors of transferring written data into the numeric controlled saws.

MiTek and Wolf have developed methods of precise plate location at each joint by projecting a laser generated image of the plate onto the timber in the manufacturing jig. These techniques are of course particularly helpful in improving the accuracy of plate placement not only on the first plate laid on the table before the timbers are laid on to it (because this particular plate could always be located by marks on the jig itself), but particularly for the top plate which is traditionally placed by the trained eye of the assembler. These techniques are still relatively new as this book goes to press, but if taken on board by the industry should significantly reduce the setting-up procedures prior to manufacturing each batch of trussed rafters, and will improve the accuracy of assembly and therefore the ultimate quality of the product.

All programs, now to varying extents, provide a 'take-off' facility for infill areas such as those around the hip and valley situations. These elements of the program are, of course, invaluable to those truss rafter manufacturers who provide 'whole roof packages', i.e. a package containing not only the roof trusses required, but also all infill timbers for hips, valleys, dormers, fascias, bargeboard and connecting metalwork.

TRAINING

All of this software sophistication is of course of little use unless the staff of each trussed rafter manufacturer are able to fully exploit it. It is here that the main differences now exist between trussed rafter manufacturers. All have the software available to them but few use all aspects of their plate supplier's software packages. Computer-aided drafting, for instance, is still not commonplace with manufacturers with the exception of those using the Wolf system, whose program seems to be particularly graphically oriented, but all use the quoting facility (although not perhaps the word processing element) (see point (3) on page 100), and all use the engineering and production data elements of the software available to them. All plate suppliers provide training for their manufacturers' staff, with MiTek and Gang-Nail generally tackling this on a planned, structured basis, offering training courses for varying levels of ability. Alpine and Wolf tend to deal with training on a more individual need basis, tuning their training to suit either their newly

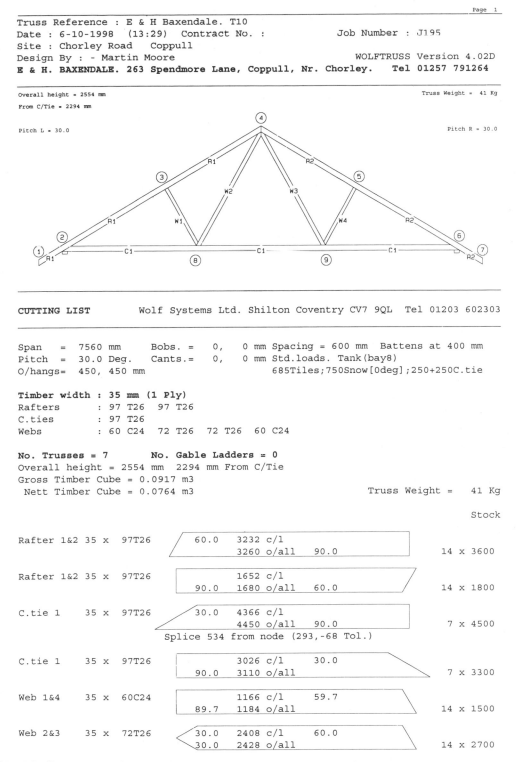

Truss Reference : E & H Baxendale. T10
Date : 6-10-1998 (13:29) Contract No. :   Job Number : J195
Site : Chorley Road   Coppull
Design By : - Martin Moore   WOLFTRUSS Version 4.02D
E & H. BAXENDALE. 263 Spendmore Lane, Coppull, Nr. Chorley.   Tel 01257 791264

Overall height = 2554 mm   Truss Weight = 41 Kg
From C/Tie = 2294 mm

Pitch L = 30.0   Pitch R = 30.0

CUTTING LIST   Wolf Systems Ltd. Shilton Coventry CV7 9QL   Tel 01203 602303

Span   = 7560 mm   Bobs. =   0,   0 mm Spacing = 600 mm Battens at 400 mm
Pitch   = 30.0 Deg. Cants.=   0,   0 mm Std.loads. Tank(bay8)
O/hangs= 450, 450 mm   685Tiles;750Snow[0deg];250+250C.tie

Timber width : 35 mm (1 Ply)
Rafters   : 97 T26   97 T26
C.ties   : 97 T26
Webs   : 60 C24   72 T26   72 T26   60 C24

No. Trusses = 7   No. Gable Ladders = 0
Overall height = 2554 mm 2294 mm From C/Tie
Gross Timber Cube = 0.0917 m3
 Nett Timber Cube = 0.0764 m3   Truss Weight =   41 Kg

                                                            Stock

Rafter 1&2 35 x 97T26   60.0   3232 c/l
                                      3260 o/all   90.0   14 x 3600

Rafter 1&2 35 x 97T26   1652 c/l
                              90.0   1680 o/all   60.0   14 x 1800

C.tie 1   35 x 97T26   30.0   4366 c/l
                                      4450 o/all   90.0   7 x 4500
                      Splice 534 from node (293,-68 Tol.)

C.tie 1   35 x 97T26   3026 c/l   30.0
                              90.0   3110 o/all   7 x 3300

Web 1&4   35 x 60C24   1166 c/l   59.7
                              89.7   1184 o/all   14 x 1500

Web 2&3   35 x 72T26   30.0   2408 c/l   60.0
                              30.0   2428 o/all   14 x 2700

Fig. 6.2 Computer printout – cutting-data.

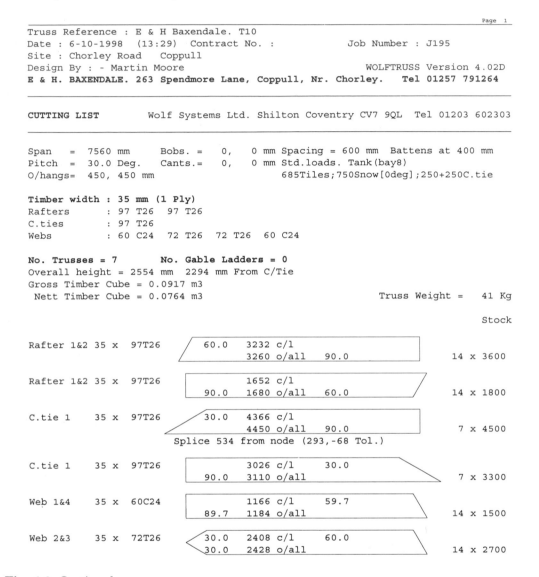

```
Truss Reference : E & H Baxendale. T10
Date : 6-10-1998  (13:29)  Contract No. :          Job Number : J195
Site : Chorley Road   Coppull
Design By : - Martin Moore                         WOLFTRUSS Version 4.02D
E & H. BAXENDALE. 263 Spendmore Lane, Coppull, Nr. Chorley.   Tel 01257 791264


CUTTING LIST        Wolf Systems Ltd. Shilton Coventry CV7 9QL  Tel 01203 602303


Span   = 7560 mm     Bobs. =  0,   0 mm Spacing = 600 mm  Battens at 400 mm
Pitch  = 30.0 Deg.   Cants.=  0,   0 mm Std.loads. Tank(bay8)
O/hangs=  450, 450 mm               685Tiles;750Snow[0deg];250+250C.tie

Timber width : 35 mm (1 Ply)
Rafters    : 97 T26   97 T26
C.ties     : 97 T26
Webs       : 60 C24   72 T26   72 T26   60 C24

No. Trusses = 7      No. Gable Ladders = 0
Overall height = 2554 mm  2294 mm From C/Tie
Gross Timber Cube = 0.0917 m3
 Nett Timber Cube = 0.0764 m3                      Truss Weight =   41 Kg

                                                             Stock

Rafter 1&2 35 x  97T26    60.0    3232 c/l
                                  3260 o/all    90.0        14 x 3600

Rafter 1&2 35 x  97T26            1652 c/l
                          90.0    1680 o/all    60.0        14 x 1800

C.tie 1    35 x  97T26    30.0    4366 c/l
                                  4450 o/all    90.0         7 x 4500
                   Splice 534 from node (293,-68 Tol.)

C.tie 1    35 x  97T26            3026 c/l      30.0
                          90.0    3110 o/all                 7 x 3300

Web 1&4    35 x  60C24            1166 c/l      59.7
                          89.7    1184 o/all               14 x 1500

Web 2&3    35 x  72T26    30.0    2408 c/l      60.0
                          30.0    2428 o/all               14 x 2700
```

Fig. 6.2 *Continued.*

acquired fabricator or the fabricator's newly acquired staff. Wolf, for instance, issue each delegate with a certificate once training is complete in the same way that the major car manufacturers ensure their engineers are trained and certificated in their particular marque and model.

TECHNICAL DATA

Each plate manufacturer produces a roofing manual to aid the trussed rafter user in the design and understanding of on-site construction of the trussed rafter roof. Both Gang-

Nail and MiTek produce bound books but it has to be said that MiTek's publication size results from some duplication of detail between the 'technical information' section and the 'information for site use' section. On the other hand the Gang-Nail book covers almost every foreseeable detail for trussed rafter roof assembly, and has a small section for the design from tables for the site infill timbers. Wolf and Alpine both produce very competent publications with Wolf's probably being the most easily read, both giving full information on trussed rafter roof assemblies. It should be emphasised that these are not design manuals but will help the architect, specifier and detailer achieve a basic roof design, which can then be engineered by the trussed rafter producer without too much alteration to the supporting structure to enable an economic roof to be designed.

METALWORK

The ability to design easily and cost-effectively sophisticated roofs has led to a greater demand for specialist metalwork to connect the roof components together. This means truss to girder, girder to girder, infill truss, infill to trussed rafter, purlin to trussed rafter and trussed rafters to wall connections. All truss nail plate manufacturers offer design advice on the correct metal fixings to be used in their roof structure, and many offer the metalwork itself via their trussed rafter manufacturers' network. The connecting metalwork from truss to truss, and truss to ancillary timber or wall plate, is more likely to be specified by the trussed rafter manufacturer who will have available to him various standard metalwork products produced by a wide range of companies independently of the truss plate manufacturers. It must be remembered that similar metalwork would be required in traditional roof construction and of course is the 'norm' in the use of steel joist hangers in timber floor construction in housing and other buildings. Of the truss plate suppliers, only Alpine issue their own builder's hardware leaflets, with Simpson Strong-Tie probably being the largest single manufacturer of specialist connectors for roof structures in this country. Figure 6.3 illustrates some of the metalwork available from Simpson Strong-Tie for roof structure assembly.

HIP ROOF CONSTRUCTION

Variation in hip construction using trussed rafters is almost endless, but the following text and illustrations give an indication of the most common forms, the illustrations being taken from the plate suppliers' technical manuals.

A common hip construction is illustrated in Fig. 6.4. This uses a flat-topped girder truss with flat-top infill trusses between girder and main roof. The girder should have been nailed (or bolted) together in the factory and is positioned on the wall plate to take the monopitch hip infill trusses it supports. The exact location of the girder may vary between suppliers but a check on the monopitch span will quickly ensure accurate location on the wall plate. The rafters of both mono, girder and infill flat-top trusses will be left much longer than necessary, these being cut off on site to fit the hip board. The off cuts can frequently be used as hip corner infill timbers where the compound cut will be precisely correct. The mono trusses may be either top- or bottom-chord supported

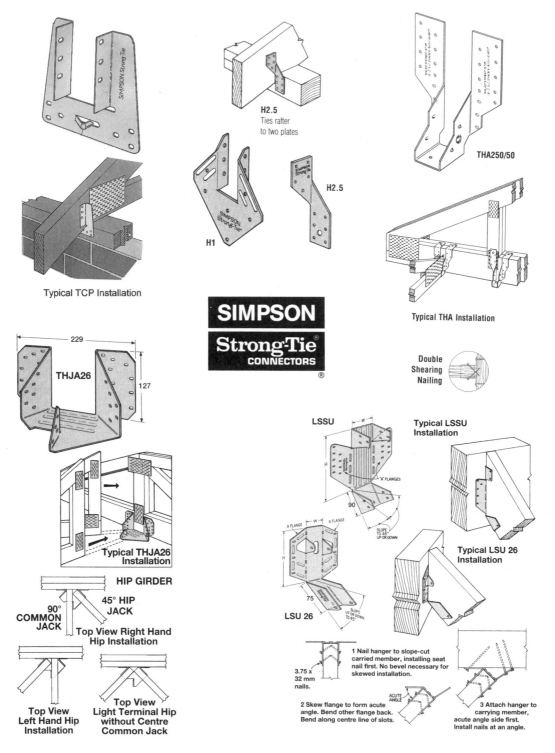

Fig. 6.3 Roofing metal work.

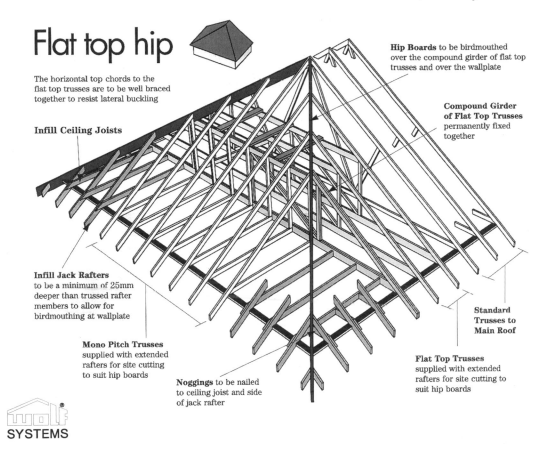

Flat top hip

The horizontal top chords to the flat top trusses are to be well braced together to resist lateral buckling

Infill Ceiling Joists

Hip Boards to be birdmouthed over the compound girder of flat top trusses and over the wallplate

Compound Girder of Flat Top Trusses permanently fixed together

Infill Jack Rafters to be a minimum of 25mm deeper than trussed rafter members to allow for birdmouthing at wallplate

Mono Pitch Trusses supplied with extended rafters for site cutting to suit hip boards

Noggings to be nailed to ceiling joist and side of jack rafter

Standard Trusses to Main Roof

Flat Top Trusses supplied with extended rafters for site cutting to suit hip boards

wolf®
SYSTEMS

Fig. 6.4 Hip roof trussed rafter layout.

and this must be clearly known before fixing to ensure truss shoes are used if needed. Figure 6.4 illustrates the above construction.

An alternative form of hip construction is to use a 'stepdown' hip construction. This utilises a similar flat-topped hip girder with oncoming monopitch trusses to form the hip end, but this time between the girder and the main ridged roof is fitted a series of simple flat-top trusses with the flat-top element increasing in height as the hip increases towards the ridge point. This saves the flying rafter or top chord of the monopitch trusses, but necessitates the fitting of noggings between these increasingly tall trusses to ensure adequate fixing for tile battens.

Other variations of hip construction exist and it is wise to seek the manufacturer's hip construction detailed notes before proceeding on site.

Bonnet or gable hip

Figure 6.5 illustrates a typical construction for this hip form. To carry the weight of the hipped element of the roof, the fink of the main roof would probably not be strong

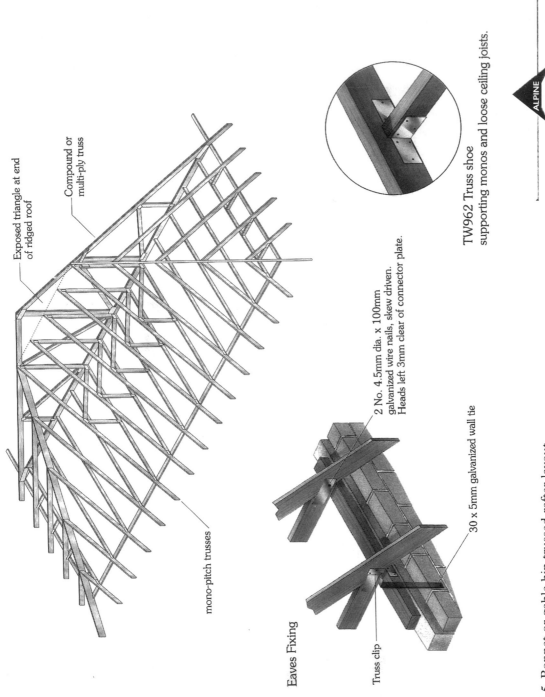

Exposed triangle at end of ridged roof

Compound or multi-ply truss

mono-pitch trusses

TW962 Truss shoe supporting monos and loose ceiling joists.

ALPINE

2 No. 4.5mm dia. x 100mm galvanized wire nails, skew driven. Heads left 3mm clear of connector plate.

30 x 5mm galvanized wall tie

Eaves Fixing

Truss clip

Fig. 6.5 Bonnet or gable hip trussed rafter layout.

enough on the bottom chord. The Howe configuration has an extra support on the bottom chord, dividing the span into four rather than three bays of the fink. Again a girder will be formed of two (or three) Howe trusses, and this time the monopitch truss will definitely be bottom-chord supported in truss shoes. The upper part of the Howe (that triangle above the monopitch roof), will have to carry some form of cladding, possibly rendering or boarding. The additional weight of this cladding must not be overlooked when designing the truss.

Barn or Dutch hip

One method of constructing this roof feature is illustrated in Fig. 6.6. This time no girders are required as there are no monopitch trusses from the hip to support. It is commonplace to use a series of flat-topped fink-based trusses, again with flying top chord to be cut to fit the small hip board needed to complete the roof feature. Infill timbers from peak of gable to these flying rafters complete the roof structure.

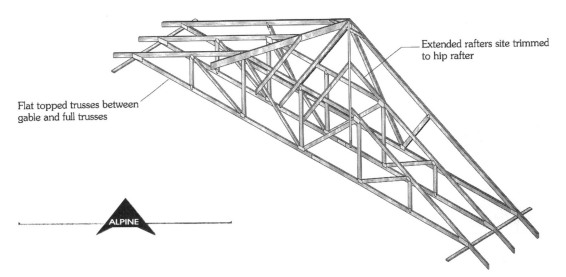

Extended rafters site trimmed to hip rafter

Flat topped trusses between gable and full trusses

Fig. 6.6 Barn or Dutch hip trussed rafter layout.

Hip corners

A common problem on many domestic buildings is the return hip. Often there are no load bearing walls on which to support the flat-top hip girder and infill trusses where they run into the adjoining roof. This necessitates a further girder of some form, as illustrated in Fig. 6.7. This particular solution is now probably the most familiar found between all trussed rafter manufacturers because in essence it is a conventional hip end construction supported at one side on the incoming roof girder truss. It is, however, a little difficult to construct on site in that the supporting compound truss has to be

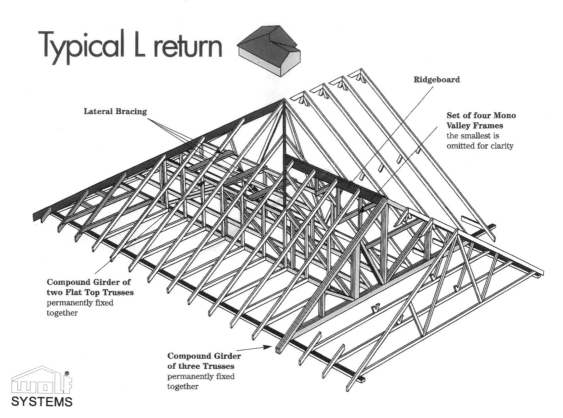

Typical L return

Lateral Bracing

Ridgeboard

Set of four Mono Valley Frames the smallest is omitted for clarity

Compound Girder of two Flat Top Trusses permanently fixed together

Compound Girder of three Trusses permanently fixed together

wolf®

SYSTEMS

Fig. 6.7 Hip and valley intersection trussed rafter layout.

effectively stabilised before it can carry the flat-top trusses it supports. This can generally only be done by using longitudinal bracing. Because the girder itself is often of different configuration (a Howe) to the simple trusses in that part of the roof (often Finks), continuity of bracing is difficult to achieve, there being few common node points. The author's preferred construction is that illustrated in Fig. 5.11, because the hip and girder have a vertical web against which the supporting Howe girder can be rigidly fixed, thus effectively stabilising both major components.

Dog leg hip/valley

For this roof form it is not technically impossible to provide a series of different shaped trussed rafters simply fanned out around the dog leg. However, this is usually prohibitive on cost as each truss would be a different shape because the span would increase while the overall height would have to remain constant, necessitating a different jig setting for each pair of trusses produced. This latter statement of course assumes both legs of the dog leg are of equal span!

The solution illustrated in Fig. 6.8 assumes the use of a girder truss at the last point into the dog leg where a truss can be placed at true right angles. This is indicated as Girder B on the illustration. A further compound truss (Girder C) is then designed to fit across the intersection of the dog leg. These girders between them carry an infill timber system to complete the two triangular infill areas.

Whilst on small spans, a compound of the fink of the common roof could be used at Girder B; the heavier load and effectively lower pitch of Girder C may dictate that some different configuration may be necessary. Obviously to gain best support for infill purlins and ceiling joist binders, the configuration of Girder B and Girder C must be identical. This invariably means a Howe girder is used in most domestic size roof constructions. The purlins and ceiling joist binders are supported at node points on heavy duty joist hangers or truss shoes, nailed or bolted to the girder truss. Similarly, a ridge must be fitted to continue the ridge line into the dog leg. On to these secondary supporting members the common rafters and ceiling joists can be fitted.

A second possible solution is indicated in Fig. 6.8 which uses a trussed purlin on which monopitch trusses can be supported. This option, of course, reduces the amount of site infill and, the author suggests, may be a more economic solution for larger spans or perhaps where the dog leg is a repetitive feature throughout the building site.

VALLEYS

Two main situations exist (see Fig. 6.9). Firstly, a load bearing wall may exist under the main roof where it adjoins the incoming roof marked 'valley/valley' on the illustration. Secondly, of course, we have to cope with no such load bearing walls in that position, this being the assumption in the illustration.

In the first situation, the main roof trusses can be supported on that wall on the 'valley/valley' line, there being no real connection needed between the two roof structures. To aid economy in production all the trussed rafters of the main roof would be made exactly the same with overhang on both sides, the overhang of the trusses in the intersecting area simply being cut off on the wall plate line as necessary. The incoming roof can be simply constructed using trussed rafters right up to that wall plate line with the valley itself being infilled in a similar way to that illustrated in Fig. 3.16, or by using prefabricated valley sets as detailed below. In either case it should be remembered that the trussed rafters, where they are not covered by tiles underneath the valley area, should be fitted with tile battens at 300 mm centres, as assumed by the structural design.

In the second situation it is clear that the main roof must be supported on a girder at the roof intersection line, the common trussed rafters of the main roof being supported in girder truss shoes as illustrated in Fig. 5.8. Figure 6.10 illustrates such a junction and explains the use of a prefabricated valley frame set. The MiTek illustration used in Fig. 6.10 highlights the problem of the eccentric load on the supporting girder, this eccentricity tending to cause the truss to roll onto the roof trusses it supports. This can be overcome by installing the details shown at the bottom of the illustration known as a torsional restraint unit for the girder. The alternative to this, of course, is the author's

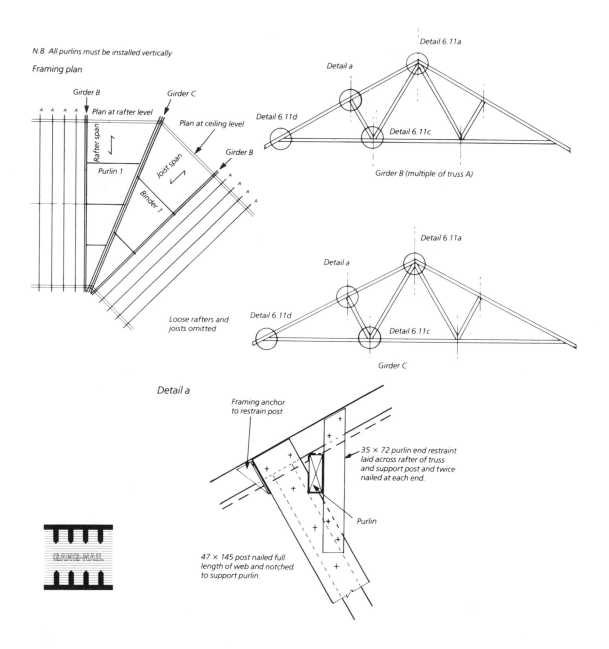

Fig. 6.8 Dog leg valley intersection trussed rafter layout.

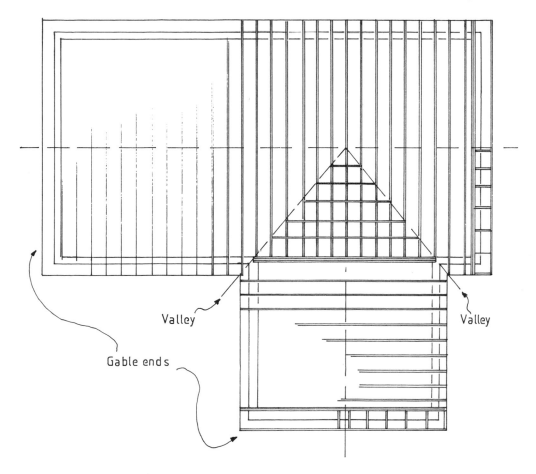

Fig. 6.9 Typical 'T' roof layout.

preferred detail indicated in Fig. 5.11 which ensures full support for the Howe girder, but would be more costly in that the truss in that area would have to be a specially constructed 'one off'.

Hip – valley intersection

We now look at a hip which overlays the valley. This presents its own special problems and cannot be solved with the designs illustrated thus far. Unlike a normal intersection, in which a full height girder can be used where the hip overlays the main roof to some degree, the depth available for the main roof truss supporting girder is drastically reduced. The solution illustrated in Fig. 6.11 uses a flat top girder truss to support both the main roof trussed rafters and the hip monos. Prefabricated frames form the valley and carry the upper part of the hip construction, these being supported off the trussed rafters. There is no major problem with the additional imposed load of the frames themselves as the principle load is from the tiles, and there is no greater area of tiling

"TEE" INTERSECTIONS AND VALLEY INFILL

The basic junction of two roofs is known as a 'Tee' intersection. Where two roof planes intersect, a valley line will be formed. The construction around the valley area is commonly formed by the use of either loose timber rafters, valleyboards and ridgeboards or by the use of pre-fabricated valley infill components (valley trusses). Fig 73 (a) (b).

Figure 73 a

Figure 73 b

VALLEY INFILL TRUSSES

It is strongly recommended that valley infill components be used in junction areas, as these provide the quickest, cheapest and most structurally effective solution to the roof framing in these areas.

The use and function of the valley infill components are more important than they appear. The individual components transfer the tile and other loadings etc, to the top chords of the underlying standard trusses in a uniform manner. And, acting with the tiling batten between each neighbouring pair of components, they provide lateral stability to the same chords. This bracing of the top chords may be chamferred to suit the pitch of the roof and skew nailed in place.

Some variations on the basic system are shown in Fig 75. Others occur from time to time and suitable layouts can be easily devised by Mitek trussed rafter suppliers.

The layboards shown in Fig 74 are in short lengths and supported off battens nailed to the sides of the rafters, to lie flush with the tops of the rafters. This procedure is time-consuming to install, but enables the felt and tiling battens to be carried through into the valley. The tile manufacturers advice should be sought to ensure correct tile and pitch suitability.

In many cases, the support for the main roof trusses may be provided by a multi-ply girder truss as shown in Fig 76 (a), with the incoming trusses supported in Hydro Air M104 Girder Truss Shoes at each position.

It is common practice on site to erect the girder truss first and position the incoming trusses afterwards.

TILING BATTEN BETWEEN NEIGHBOURING VALLEY INFILL TRUSSES

TEE INTERSECTIONS

GIRDER TRUSS

Figure 74

VALLEY INFILL TRUSSES

LAYBOARDS   RETURN SPAN STANDARD TRUSSES

Figure 75

Please note that in cases where both the girder span and the incoming truss span is large, the torsional restraint detail shown in Fig 76 (b) should be used.

As described above, the valley construction should include intermediate tiling battens between neighbouring valley infill trusses, to provide the correct restraint for the rafters of the underlying trusses.

Suggested further reading Sections: 1.5, 3.6, 3.9

TYPICAL GIRDER TRUSSES

Figure 76 a

TYPE 800   TYPE 1200

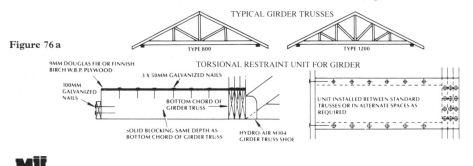

9MM DOUGLAS FIR OR FINNISH BIRCH W.B.P. PLYWOOD

TORSIONAL RESTRAINT UNIT FOR GIRDER

3 X 50MM GALVANIZED NAILS

100MM GALVANIZED NAILS

BOTTOM CHORD OF GIRDER TRUSS

UNIT INSTALLED BETWEEN STANDARD TRUSSES OR IN ALTERNATE SPACES AS REQUIRED

SOLID BLOCKING SAME DEPTH AS BOTTOM CHORD OF GIRDER TRUSS

HYDRO-AIR M104 GIRDER TRUSS SHOE

MII
MiTek Industries Ltd

Fig. 6.10 Valley intersections trussed rafter roofs.

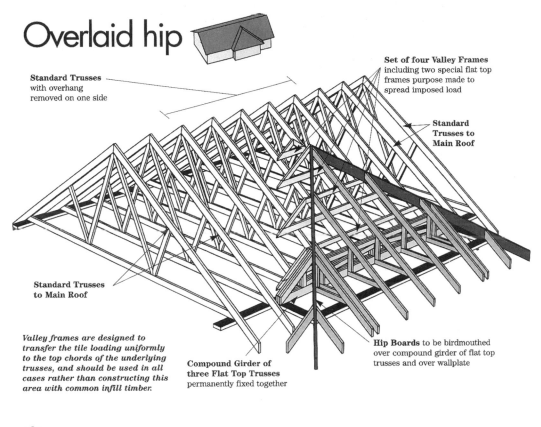

Overlaid hip

Standard Trusses with overhang removed on one side

Set of four Valley Frames including two special flat top frames purpose made to spread imposed load

Standard Trusses to Main Roof

Standard Trusses to Main Roof

Valley frames are designed to transfer the tile loading uniformly to the top chords of the underlying trusses, and should be used in all cases rather than constructing this area with common infill timber.

Compound Girder of three Flat Top Trusses permanently fixed together

Hip Boards to be birdmouthed over compound girder of flat top trusses and over wallplate

SYSTEMS

Fig. 6.11 Overlaid hip–valley intersection.

than there would be had the hip not been imposed. The hip would be completed with a short length of ridge board running from the peak of the hip to the main roof itself. This hip board could either be supported on a trussed rafter if positioned in line with the peak of the ridge of the hip, or on the valley lay boards themselves supported by a trimmer between the trussed rafters at that point. These latter items have been omitted from this particular illustration for clarity, but similar details can be found in Chapter 3 on traditional roof construction.

ATTIC TRUSSES

Sophisticated computer programs again help to analyse the complex structure of the attic, a truss form (unlike most others) which is not fully triangulated in that part which is the attic room itself. Because of the higher bending loads which occur in the sloping ceiling section of the rafters and the floor joists, larger timber sections are required. These are usually 47 mm thick and often 200 or 225 mm deep. The weight of the truss also becomes a significant factor when designing, manufacturing, handling and fixing such trusses on the building. The overall height of the truss is another consideration. A 7 m span, 45° pitch, for instance, will be over 3.6 m high plus the overhangs. This will present transport problems and be difficult to handle on site without mechanical assistance. To overcome the height problem, the attic is often split to give the 'top-hat' design referred to in Fig. 5.19. All truss plate system manufacturers have similar solutions to attic roof constructions.

Innovation

The ability to design larger spans for attic trusses without internal support has been made possible by the development of a method for localised stiffening of the floor joist or the sloping ceiling rafter. The method involves plating two pieces of timber edge-to-edge within the truss to stiffen the highly stressed areas; this avoids having to use larger timbers throughout the truss or resort to the use of 'scabs', bolted or nailed to the side of the truss top chords. All plate system manufacturers employ this technique, variously called: *Superchord* (Gang-Nail), *Stackchord* (MiTek), *Wolfchord* (Wolf Systems) and *Twinachord* (Alpine).

There are undoubtedly many advantages:

- *Saving in timber volume:* saving in truss weight and cost.
- *Saving in timber thickness:* more trusses on the lorry load and the use of standard truss clips.
- *Saving in not having to fit scabs:* only one thickness to be birdsmouthed on site rather than truss plus scabs. Automatically increases the depth of the rafter sloping-ceiling section, thus enabling easy accommodation of insulation and ventilation (see Fig. 11.7).

Figure 6.12 illustrates Gang-Nail's 'Superchord' plating for various uses.

RAISED TIE TRUSSES

Refer to Fig. 5.23I.

The extended rafter sections clearly carry the whole of one side of the truss, necessitating a larger section of timber for the sloping ceiling length. All plate producers have three alternative solutions:

(1) Increase the cross section of the top chord over the whole of the top chord length; if this can be done by using 36 mm stock sections then there is no major cost implication. If, however, it necessitates the use of 47 mm timber then clearly all the

Superchord Plating

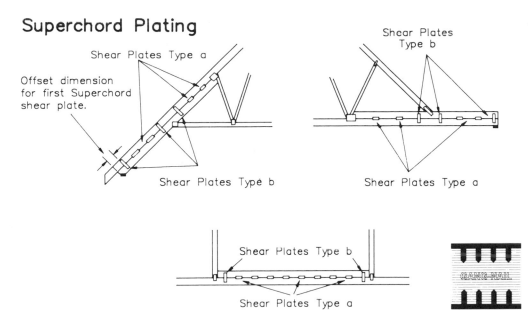

Fig. 6.12 Superchord – typical uses.

truss members must now be 47 mm, and there is a significant cost implication brought about purely by the need to overcome a structural problem occurring in the sloping ceiling area of the truss.

(2) Add scabs (i.e. additional pieces of timber) nailed or bolted to each side of the rafter from the wallplate to the first node joint. This at least leaves the truss itself using relatively lightweight timbers, but it does mean that the bird's mouth becomes a difficult joint to cut on site.

(3) Use the plated chord discussed in 'Innovation', above (see Fig. 6.13), this time taken from Wolf's technical manual. This particular detail allows a greater length of sloping rafter, i.e. a greater raise to the tie than is generally practical with methods 1 and 2, above.

The raised tie roof truss imposes a horizontal load onto the wall plate and wall when loaded. Unlike the conventional truss which is fully triangulated by virtue of its bottom chord fixing the top chords together, the raised tie roof completes its triangulation at the flat ceiling to sloping ceiling intersection. The extended rafters then act as a beam which deflects to give an element of lateral movement. We are now back in effect to a collar roof (see Fig. 1.2).

To overcome at least the horizontal thrust from the roof structure and its imposed loads, a sliding truss clip or 'glide shoe' can be used. These are referred to in Fig. 6.13 and illustrated in Fig. 6.14. The shoe consists of a flat galvanised steel plate which is fixed to the underside of the rafter seating where it bears on the wall plate. A special

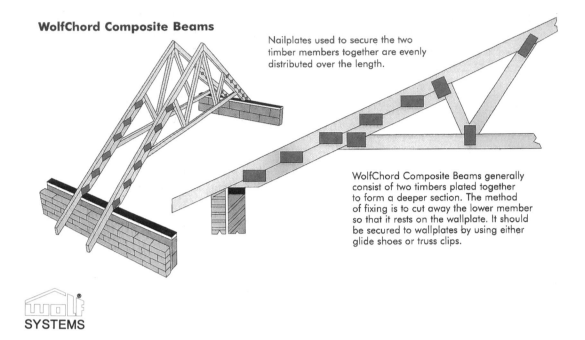

WolfChord Composite Beams

Nailplates used to secure the two
timber members together are evenly
distributed over the length.

WolfChord Composite Beams generally
consist of two timbers plated together
to form a deeper section. The method
of fixing is to cut away the lower member
so that it rests on the wallplate. It should
be secured to wallplates by using either
glide shoes or truss clips.

wolf
SYSTEMS

Fig. 6.13 Raised tie bracing.

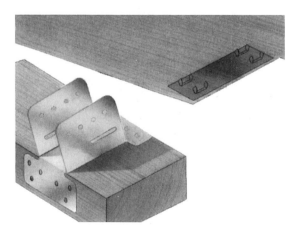

TW964 GLIDE SHOE

ALPINE

Fig. 6.14 Truss glide shoe.

truss clip is then fitted to the wall plate and the truss lowered onto it. The Alpine TW964 glide shoe illustrated can accommodate up to 15 mm of horizontal movement, but Alpine recommend this is limited to 6 mm on each bearing within the truss design. To allow the shoe to take up its movement, nails are fixed in the slots but not fully driven home, thus allowing the truss to slide as the roof is constructed and loaded. When the roof is fully loaded (i.e. tiled), further nails are added to fix the truss in its deflected working condition and the first-fix nails are fully driven home. The glide shoe no longer 'glides', acting then as a conventional truss clip.

Bracing the raised tie in the standard configuration truss area follows normal procedures but presents its own problems in the sloping ceiling area where special attention must be paid to lateral stability, i.e. avoiding the domino effect of the trussed rafters. Options 1 and 2, above, produce relatively stiff trusses in this area and the length of the sloping ceiling is relatively small compared to the plated chord solution. This latter technique needs special considerations, as follows:

(1) Because the thickness of the plated chord is relatively slim compared to its depth, buckling must be avoided and solid blocking should be fitted between the trusses in the raised tie area. This would follow similar practice used for floor joists.
(2) Because the length of the sloping ceiling rafter is probably greater, there is probably a greater need to diagonally brace this element of the roof.

This additional bracing can be provided in three ways:

(1) Sheath the roof on the top of the rafters, a common practice in Scotland and Scandinavian countries where it is known as a sarking (see Fig. 5.22).
(2) Continue the under rafter diagonal bracing down over the underside of the sloping ceiling rafter. The change of plane presents problems which have to be effectively dealt with by blocking down from the common diagonal brace to the plated chord diagonal brace. This also means further packing out for fitting of plasterboard.
(3) Fit ply between rafters utilising the solid blocking mentioned above as support. Figure 6.15, again from Gang-Nail's Technical Bulletin 126, shows the use of plywood in this situation.

Option 3 is preferred by the author because the ply can perform two functions: firstly it stabilises the roof and uses the solid blocking as support; secondly it provides the ventilation void in the sloping section. This is basically the same technique as that used for stabilising attic trusses (see Fig. 7.22).

PUNCHED NAIL PLATE JOISTS, RAFTERS AND PURLINS

Initially developed in the USA and Canada for long span floor and flat roof joists, the product consists of an upper and lower timber chord spaced apart by special metal 'V' pressings which at the top and bottom of the 'V' have punched nails similar to the

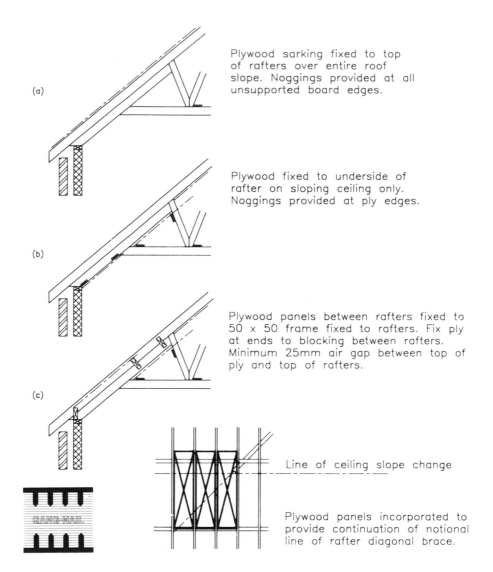

(a) Plywood sarking fixed to top of rafters over entire roof slope. Noggings provided at all unsupported board edges.

(b) Plywood fixed to underside of rafter on sloping ceiling only. Noggings provided at ply edges.

(c) Plywood panels between rafters fixed to 50 x 50 frame fixed to rafters. Fix ply at ends to blocking between rafters. Minimum 25mm air gap between top of ply and top of rafters.

Line of ceiling slope change

Plywood panels incorporated to provide continuation of notional line of rafter diagonal brace.

Fig. 6.15 Raised tie bracing.

trussed rafter connector plate. When pressed into the top and bottom flanges or chords, these 'V's produce a lattice girder.

The advantages:

(1) The economic use of timber.
(2) A precision engineered component.
(3) Easy installation of services such as electric cables, plumbing pipes, and small ducts through the lattice.
(4) Wide fixing chords for floor and ceiling decking.

(5) The same component can be used for joists, rafters, purlins, etc.

(6) No on-site waste because the component is manufactured to a precise requirement.

(7) Can be designed to span clear between external walls without internal load bearing partitions and their associated costs of additional foundations.

(8) Can be designed for top or bottom chords support.

(9) A considerably improved product weight/span ratio compared to solid timber.

One of the first truss system manufacturers to produce a similar product was Gang-Nail who many years ago produced a product called 'Econoflor'. This earlier version of the 'V' lattice beam used timber lattice members connected to the top and bottom chords with the same punched metal connector plates as used for trussed rafters. The labour content in cutting the numerous small lattice timbers and indeed placing them in special jigs made the product less than economically viable compared to conventional solid timber, and at that stage Gang-Nail had problems in convincing the authorities of its performance in fire. Gang-Nail have now reintroduced this construction under the name 'Eco-floor'. They claim it is now a viable product compared to the alternative forms of manufactured joist where the floor zone is limited to approximately 200 mm.

Gang-Nail have since produced a new product called Space Joist, illustrated in Fig. 6.16, using the metal 'V' nail plate construction. MiTek have a similar product called

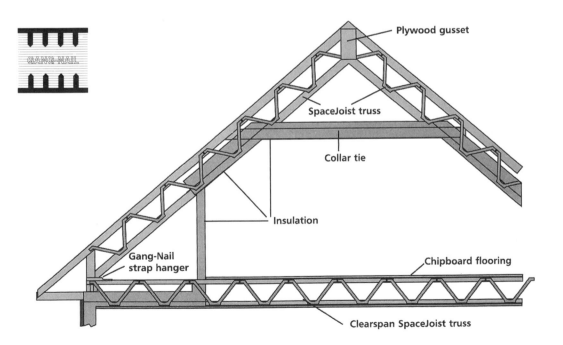

Fig. 6.16 SpaceJoist attic.

Posi-Strut; this is illustrated in Fig. 6.17 showing a typical joint detail for attic roof construction using the Posi-Strut for floor joist and rafters.

Trimming for openings can be simply done as illustrated in Fig. 6.17. MiTek concentrate on the production of Posi-Strut with a few of their fabricators who specialise in the product, investing in the dedicated jigs required, thus making them highly efficient and improving the cost effectiveness of the component.

One of the initial objections to the use of lattice joists as mentioned above is the spread of fire through the floor or rafter void. Testing has been done by both Gang-Nail and MiTek who have developed approved constructions for up to one hour fire resistance for use as floor joists, thus clearing obstacles from the path of those wishing to specify this relatively new building product.

ROOFING   FLOORING

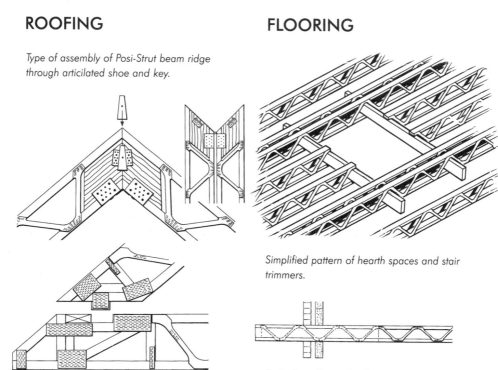

Type of assembly of Posi-Strut beam ridge through articilated shoe and key.

Simplified pattern of hearth spaces and stair trimmers.

Typical cantilever detail.

Type of assembly of truss foot: Posi-Strut with the principal rafter on roof beam and return of load moving horizontally.

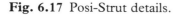
MiTek Industries Ltd.

Fig. 6.17 Posi-Strut details.

FUTURE DEVELOPMENTS

The last edition of this book identified the ever advancing sophistication of computer programs on offer to the trussed rafter manufacturers. The continuing need to be competitive will drive the plate system owners on to further advances both in the degree of roof shape complexity possible to process and in the simplicity of operating the programs produced.

Mi-Tek and Wolf now have available laser projection equipment which as this book goes to press is in limited use but will undoubtedly become more widespread in the future. Doubtless this type of feature will be further developed and in the next edition of this book we may well be reporting on the use of robotics in the selection and placement of the plates and the timber components of a trussed rafter. Systems already exist to program the actual pressing machine to stop and press the plates at each node point on the truss. With the programming time necessary to achieve this degree of control, and therefore the cost of input, falling all the time, the economics of using this technology on relatively small batches of trusses becomes viable.

Eurocode 5 is now upon us and over the next few years British Standard 6268 will gradually change to be its equivalent. The basis on which timber structural design is undertaken in the UK will change as Europe moves towards an integrated method based on limit state design.

Finger jointed timber for trussed rafter construction has not become popular, presumably for economic reasons, but the process is structurally sound and when the economic balance between finger jointing and the splice joint reverses, things may well change. This is likely to be driven by fabricators seeking cost reductions because it must be remembered that those responsible for designing the computer programs, that is the nail plate manufacturers, must have a vested interest in keeping the splice plate in use.

Customer and architectural requirements will push forward trussed rafter design to achieve their aesthetic ends, thus advancing the boundaries of trussed rafter roofs and making them even more common in the future for non-domestic structures.

CHAPTER 7
Roof Construction Detailing

GENERAL

This chapter contains a number of construction details which relate to all types of roof construction and to avoid repetition in Chapters 2, 4 and 5, they have been grouped together in this reference chapter.

The items included are as follows:

STORAGE AND HANDLING OF TIMBER AND TIMBER COMPONENTS

Timber is one of the oldest building materials known to man and is still the least respected of all on most building sites. It is probably its resilience to misuse which allows bad site practices in the storage and handling of timber to be tolerated. Generally it is only those timbers which are seen in the finished building which are afforded some respect and protection, whilst the often unseen structural timbers are frequently left unprotected and poorly stacked. Stress graded floor joists are not infrequently used as scaffold boards or barrow runs, and after having successfully survived these temporary functions are then built into the property.

The practicalities of construction work make it almost impossible to protect timbers completely, immediately after they are fixed on the house, but rapid enclosure of the roof structure and good ventilation will do much to prevent subsequent problems resulting from shrinkage on drying out.

Deliveries

It is not uncommon for the so-called carcassing timbers to arrive on site in a damp or even wet condition from the timber merchant for indeed most carcass timbers are not stored under cover. Such timber should be returned to the merchant for dry stock. If the builder has to accept wet timber, he must immediately take steps to achieve as much drying out as possible, before building the timber or components into his house.

Structural timbers such as rafters, purlins, ceiling ties, floor joists, etc., will often be delivered strapped with steel or nylon bands in house sets and may well be off-loaded by a lorry mounted crane or fork-lift truck. If the timber is at all damp then the straps should be split immediately, the timber inspected for quality and the consignment checked against the delivery note and the cutting schedule for the timbers required. Any shortages should then immediately be notified and rectified before construction starts.

The timber should be restacked on sets of bearers ideally 150 mm minimum off the ground and with thin sticks placed vertically in line with the bearers between each layer of timber to facilitate rapid drying. Any subsequent packs of timber placed on top of this stack should have their bearers again directly in line with the bearers of the pack below to avoid distortion of the timber, which could result in difficulties in fixing and certainly in poor quality in the completed building. Figure 7.1 illustrates a poorly stacked set of timber.

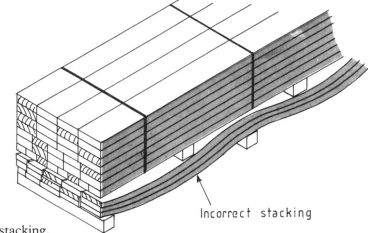

Incorrect stacking

Fig. 7.1 Timber stacking.

Protection

Wet or dry timber should be covered loosely with waterproof sheeting, leaving the ends of the packs open for air to flow through the stack. This will avoid sweating within the pack and the possibility of resulting mould growth should the timbers be left in store for some time. If covered storage is available use it, even if moving it into store takes longer than leaving it where the lorry dropped it. Dry timbers are both lighter to handle and easier to work, and most importantly they present fewer maintenance problems in the completed building because of the reduced degree of shrinkage.

Manufactured structural timber components

Extra care is needed with the handling and storage of manufactured structural timber components. Rough handling can cause structural damage which may not become apparent until the component is built in and loaded. The cost of replacement at that stage can be very high, and of course can result in delays to the building process.

Long purlin beams of solid, laminated, or ply box or web construction should be lifted from the delivery vehicle upright and not flat (see Fig. 7.2). Webbing slings from a spreader beam should be used in conjunction with either a crane, fork truck or forks on a digger and then placed on level sets of bearers and protected as described above. Major structural items should be delivered to the site as close to the actual building-in requirement as possible, thus avoiding storage. Major purlin beams, for instance, should ideally be off-loaded from the delivery vehicle by crane and hoisted directly into position in one continuous operation. This of course avoids the costs of double handling and the problems associated with storage on a restricted building site. If storage is unavoidable, then prepare the area to receive the components *before* delivery: this will speed off-loading and again save significant costs.

There has long been concern over the unsatisfactory handling, storage and protection of truss rafters on many building sites. Many operatives do not appreciate the relative flexibility and fragility of the trussed rafter when being carried in its horizontal state, rather than in its vertical state in which, of course, it is designed to carry its working loads. The NHBC, in its standards on pitched roofs 7.2 in clause S4, sets out the correct storage and handling of the components, and the Trussed Rafter Association deals with this in more depth in its technical handbook. As this book goes to press a new technical handbook has been published and is more detailed than the 1997 version. British Standard 5268: Part 3 sets out standards in section 9 for handling, transportation and storage.

Most trussed rafter manufacturers deliver the components on special vehicles on which the trusses are stacked vertically – they should remain that way on site during handling and, of course, in their working position. It is the only way in which they can be considered to be a self supporting component. Flat stacking, although sometimes unavoidable, is not in the writer's opinion considered satisfactory, and certainly requires far more attention to correct preparation of the storage area on site.

Preparation of the store area

Trussed rafter manufacturers invariably deliver the components to an agreed time schedule to the builder. For this reason there is no excuse for the site staff not being ready to receive the components. Preparation of the storage area, particularly on larger

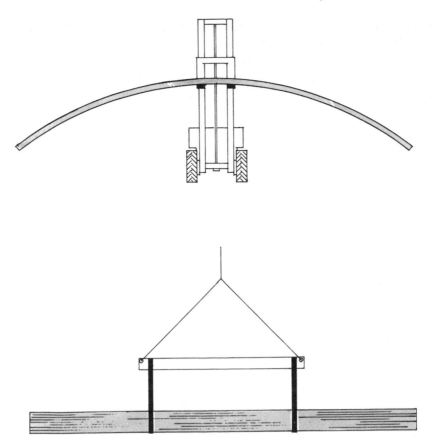

Fig. 7.2 Lifting large and long purlins.

building sites, should be carried out well before delivery, and on the very largest sites a permanent truss storage system built as part of the site set-up (see Fig. 7.3).

A simple scaffolding frame is all that is required to support this relatively light component, with main bearers placed at approximately wall plate position and possibly further bearers placed at approximately bottom chord third point positions. It is fully appreciated that more than one truss span is required on any one individual site, but careful consideration of the plans will give a satisfactory set of support positions. Having set out the horizontal supports, which should be at least 300 mm off the ground, a triangulated vertical support system should be placed at the back of the intended stack, giving support to the trussed rafters at approximately their quarter span points. This back support should be inclined backwards away from the horizontal bearers to prevent individual trusses falling forward and causing damage.

The delivery

On arrival of the delivery vehicle check the trussed rafters against the delivery ticket and the materials requisition schedules or plans before off-loading the lorry. Assuming that all is satisfactory off-loading can commence, giving the trussed rafters a physical inspection as unloading proceeds.

Many trussed rafter manufacturers strap their trussed rafters in packs of ten or twelve and if mechanical handling is used on site the rafters are best left in their strapped packs, at least until placed in the storage rack. A spreader beam as described above should be used with webbing slings passed through the quarter points on the top chord or at some similar node point. The pack should be stabilised whilst lifting, either by direct hand control if at ground level or, if the pack is to be lifted straight onto the roof plate, then by ropes attached to the heel joint positions. Strapped packs should then have the bands cut whilst they are in store on site, because the timber used in trussed rafter construction is invariably quite dry and any moisture uptake on site, whether the stack is well protected or not, will cause swelling of the timber resulting in the bands crushing those trussed rafters on the outside of the pack. Where mechanical handling is not available then obviously the straps must be cut on the lorry and the trusses manhandled on to their storage racks.

Again it must be emphasised that the trussed rafters must be carried *vertically* not horizontally. Horizontal transportation, frequently with one man at a position about one quarter way up the rafter, will result in serious 'whipping' of the trussed rafter and can cause plate disturbance and therefore damage to the component. Some trussed rafter manufacturers occasionally deliver the trussed rafters inverted for transport economy reasons. These should be off-loaded in their inverted position, carefully laid on the ground and then the peak lifted into a vertical position whilst the bottom chord is still on the ground. The truss should not be flipped over whilst being supported at the heel joints.

Figures 7.3–7.5 inclusive illustrate some of the points discussed above.

Protection

When the delivery has been safely stored in the racks, plastic sheeting protection should be fitted over the top chords at least, and loosely over the ends of the package. Care must be taken to maintain adequate through flow of air to prevent sweating within the stack.

Once the trussed rafters are constructed into their roof form, enclosure of that roof should be as fast as possible. At least felt and battens should be applied to prevent unnecessary wetting of the trussed rafter and, of course, the building structure below.

Man-power

Trussed rafters are very light components and for this reason are generally easily handled by two men. Roof spans above 9 m and pitches above 35°, however, result in a very awkward frame to handle and additional assistance will be required. British Standard 5268: Part 3 clearly states that trussed rafters forming girder trusses should be fabricated into their girder at the factory. This is now generally offered by the trussed rafter manufacturer to whom handling by fork trucks presents no problems, but once

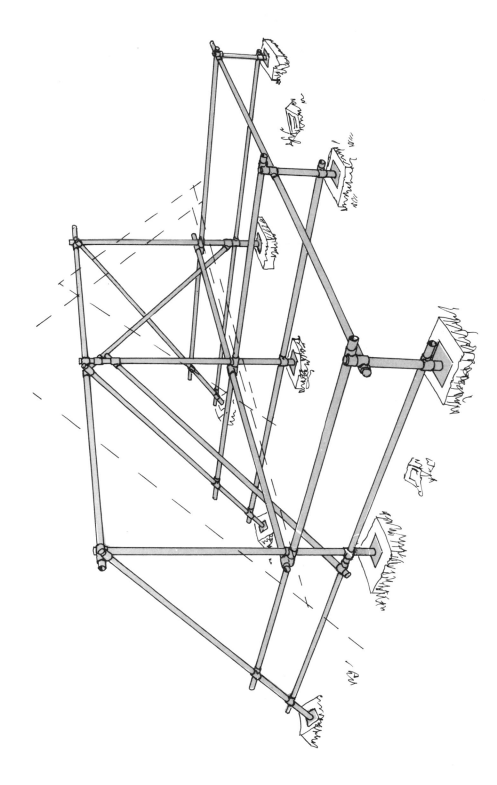

Fig. 7.3 Trussed rafter storage frame.

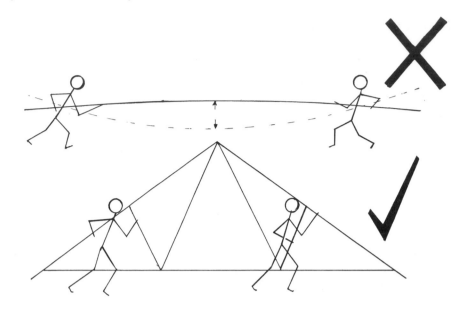

Fig. 7.4 Manhandling trussed rafters.

the girder arrives on site it may take six men to unload. The problem still remains of lifting the girder onto the building – clearly a crane should be considered to ensure safe handling of this large major structural component and to safeguard operatives from lifting excessive loads.

Similarly, the attic truss fabricated, generally from 47 mm timber and up to 3.9 m tall, presents problems; it is heavy and very unwieldy by virtue of its height. Cranes are, in the author's opinion, essential in handling large attics and attic girders, and delivery of the components should be organised with the manufacturer to ensure crane off-loading is available to lift the components, preferably directly onto the wall plates.

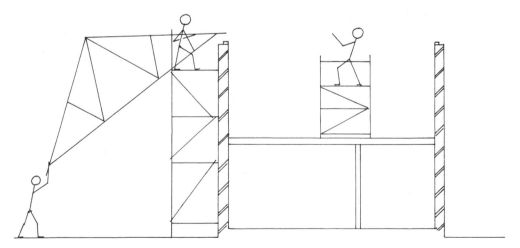

Fig. 7.5 Lifting trussed rafter on to roof.

PRESERVATIVE TREATMENT

Timber maintained at a moisture content below 20% is unlikely to be attacked by timber decaying fungi. However, the same does not apply to insect attack. Building legislation does not at present require structural timbers to be preservative treated (except those in flat roof constructions), with the exception of an area of England affected by the house long horn beetle. This area is defined in the Building Regulations. NHBC require external cladding to be treated with preservatives; this includes fascias, barge boards and timber soffits.

The various types of preservative and their application processes are set out in BS 5268: Part 5. It is not intended to enter into the detail of timber treatments. Suffice it to say that two basic types are used for constructional timbers in this country, the first being a waterborne preservative system, the second being organic solvent-borne. Their application can be by total immersion or vacuum and/or pressure depending on the type used. Brushing of these preservatives is generally not adequate to afford a high degree of protection for structural work.

Costs

The cost of timber treatment when taken in context with the overall cost of building a house is minimal, and it is therefore to be strongly recommended whether required by the regulations or not. For the traditional cut roof or those timbers site fitted to the other types of roof discussed in this book, the additional cost is approximately 20% of that of the raw material. With trussed rafters the additional cost varies between approximately 5 and 10%, depending upon the size and complexity of the trussed rafter in question.

With the increased emphasis on high insulation values, the roof structure is becoming increasingly cold. Although the roof is now usually well ventilated the cold metal fixings, be they nail plates, truss clips or gable straps, can attract condensation which could clearly be absorbed by the timber to which they are fixed. Preservative treated timber will obviously resist any degradation which could occur. Ventilation has of course slightly increased the risk of insect attack in roof spaces, although incidence of this is extremely low. Experience with well ventilated roofs is relatively limited, however, the regulations requiring ventilation having been introduced only in the past few years. For this reason again it seems prudent to preserve the roof structure.

A question which had caused some concern was the possible corrosion of the galvanised punched metal plates used on trussed rafters, when they are used in conjunction with the water-borne 'CCA' type preservatives. The concern was that the chemicals used in the treatment could cause the zinc in the galvanising to be oxidised, thus leaving the mild steel unprotected with the resulting risk of rusting. The Building Research Establishment information paper IP 14/83, referred to earlier, contains comment on a corrosion survey carried out on trussed rafter roofs. This states 'Thus whilst CCA does lead to more corrosion, this is still at a low level and on present evidence is unlikely to be of any structural significance'.

Particular care is needed with CCA preservatives to ensure that the timber is dried out thoroughly before the metal plates are fixed at the factory (BS 5268: Part 3: 1998, section 5, paragraph 5.5 deals with preservation treatment of trussed rafters). This significantly reduces the risk of plate corrosion by the preservative. It must be said, however, that the practicalities of carrying out this drying at the factory, bearing in mind the relatively short notice given to the manufacturer to produce the trusses, is impractical. It is therefore strongly recommended that CCA preservatives are not used with trussed rafter assemblies. The specification should therefore call for organic solvent based or the new generation of waterborne organic wood preservatives.

In situations of high hazard in the roof space, e.g. over a swimming pool, then preservation should be considered essential. Trussed rafters used in these conditions should use stainless steel plates, and either type of treatment system would then be acceptable. Other conditions of high hazard are, in the writer's opinion, those areas of the country immediately adjacent to coastlines, where the corrosive salt-laden atmosphere is of course encouraged to enter the building via the ventilating system.

Preservative identification and safe handling

When preservation has been specified, there is obviously a responsibility on the part of the contractor to make sure that the specification has been met. Many of the organic solvent based preservatives carry a faint tracer dye (often light reddish brown) but some manufacturers use a clear preservative. With COSHH regulations now in force, much of the telltale oily smell of organic solvent preservative will have evaporated by the time the trusses arrive on site. These regulations require safe handling of preservative treated timber to protect operatives from the chemicals contained in the preservatives themselves. This means a delay of some 48 hours between treatment and handling the timber to manufacture the trussed rafter. Some manufacturers overcome this delay by pretreating all timber stock, but then cut ends exposing untreated timber have to be swabbed with preservative before assembly. If specifying treated timber trussed rafters the purchaser should be aware of a slightly longer delivery time required by some manufacturers.

Identification of organic solvent treated timber may be possible by smelling the trussed rafters, especially towards the middle of the pack as they are delivered on the vehicle; these trusses will not have evaporated quite so much solvent and will give the distinctive oily smell of the organic solvent. Alternatively, a sliver of timber cut from the corner of one of the truss components exposing a fresh surface may indicate treatment. Finally, if in doubt, simple preservative testing kits are available from most of the preservative manufacturers, if any doubt exists. For security it is best to specify that the trussed rafter manufacturer provides a signed certificate of treatment for the trusses in any one consignment.

The CCA preservatives, being waterborne, have little or no odour. Identification is generally easily made by the light green colour imparted to the timber by the chemicals involved. If any doubt exists, simple tests can be carried out, details of which are available from the manufacturers of the treatment system or from the treatment plant used. Again, for security, a certificate of treatment should be obtained and for the safety of operatives the manufacturer's 'safe handling' instructions should be available both at the factory and on site.

WALL PLATES AND FIXINGS

The wall plate is the foundation for the roof. Care in setting out and bedding the plate will not only make the roof construction easier, but will result in a more sound construction. Poor alignment, in particular with the purlin in a traditional roof, will mean that each common rafter has to be individually fitted, making precutting of the birdsmouthing to a master pattern impossible. The additional time and cost can be considerable and again can cause delays to the construction process. Figure 7.6 illustrates the effect on a rafter of poor alignment.

Recommendations on wall plates vary bearing in mind there may well be good reasons for the different specifications depending on whether a trussed rafter or traditionally cut roof is being constructed. NHBC, in its publication *NHB Standard – Pitched Roofs* 7.2, under the heading 'traditional cut roofs' stipulates a minimum of 100 mm width × 38 mm depth for Northern Ireland and the Isle of Man, 100 mm width × 25 mm depth for Scotland, and 75 mm minimum width × 50 mm thickness for other areas of the UK. British Standard 5268: Part 3: 1998, clause 7.3, stipulates that as a guide the bearing length of the trussed rafter should be not less than 0.008 times the span of the trussed rafter with a minimum of 75 mm unless design calculations show otherwise. Wane should not be permitted. Where wall plates are bedded onto masonry walls a minimum thickness of 47 mm applies.

Wall plates should be half lapped at junctions as indicated in Fig. 7.7 and also where

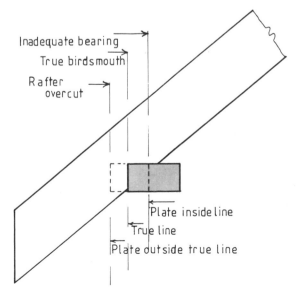

Fig. 7.6 Wall plate alignment.

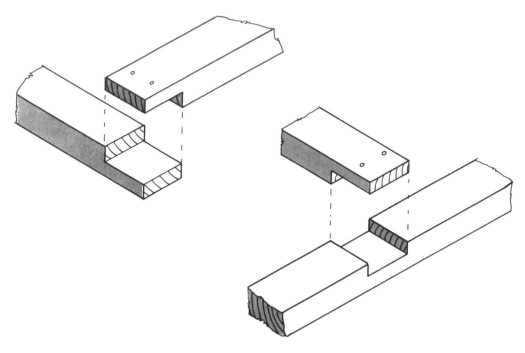

Fig. 7.7 Wall plate joints.

they butt in their length. NHBC stipulates that wall plate timbers should not be less than 3 m in length.

Strapping

Where the roof is in an exposed location, or the tiles or roof covering are very light in weight, the roof designer may require the wall plate to be strapped down to the supporting walls below. The straps used are standard components readily available from builders' merchants. Two main types of straps exist, firstly that built-in to the course of inner skin blockwork and secondly that screwed (NHBC require a minimum of three screws) to the inner skin blockwork. Both are securely fixed to the plate usually by nailing (see Fig. 7.8).

On timber framed housing the plate may be adequately secured to the frame of the house by nailing at the centres and to the specification laid down in the nailing schedule provided by the house designer. If the timber frame method uses a separate binder or wall plate across the top of the panels, this needs to be half lapped in relation to the panels to which it is fixed. This detail is shown in Fig. 7.9.

Fixing the roof to the plate

Having secured the plate, the timbers, trusses or trussed rafters need securing to the wall plate itself. This can be done by skew nailing on all forms of roof, but on trusses or

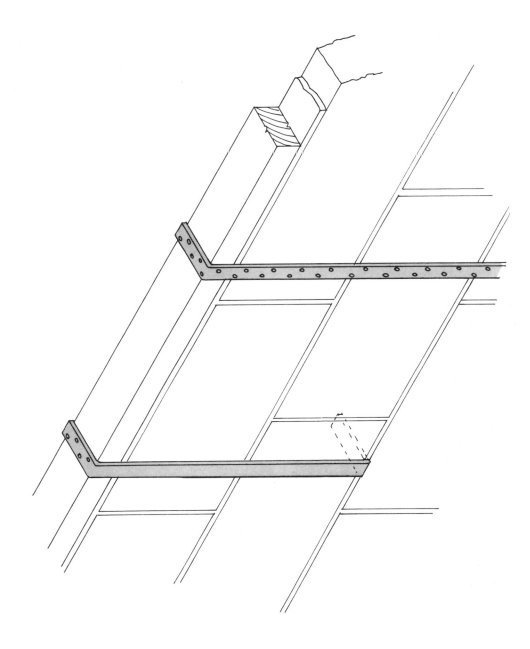

Fig. 7.8 Wall plate straps.

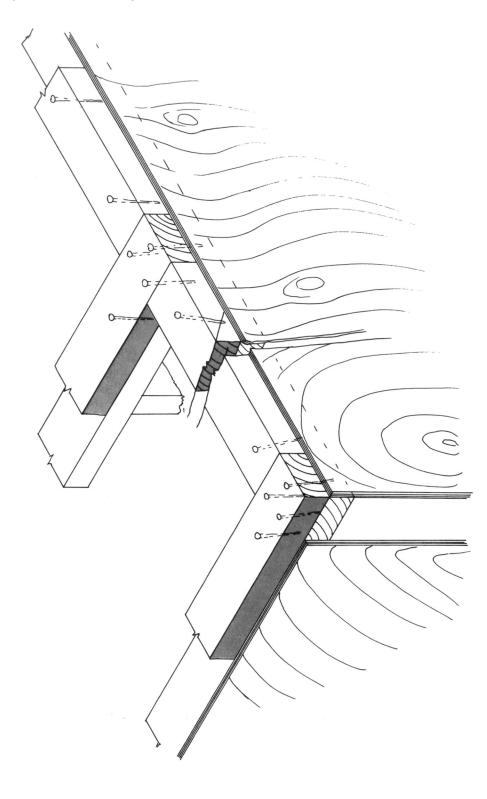

Fig. 7.9 Wall plate – timber frame construction.

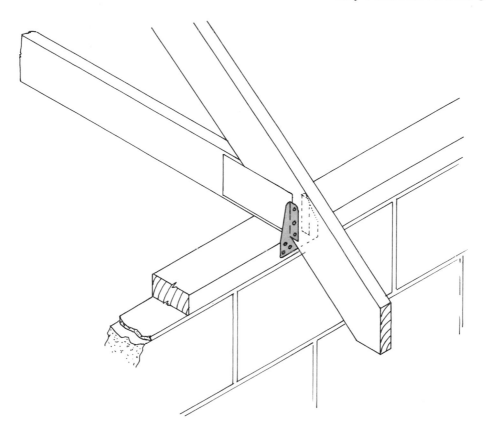

Fig. 7.10 Trussed rafter wall plate clip.

trussed rafters there is a danger of the skew nails disturbing the joint, or themselves being deflected by the plates or connectors, resulting in an ineffective joint. Truss clips or framing anchors should be used as illustrated in Fig. 7.10, and these again are readily available items from the builders' merchant. Alternatively, and particularly on the bolt and connector roof, the truss may be strapped directly to the wall below, as indicated in Fig. 7.11.

It is *strongly recommended* that for trussed rafter roofs the truss clip is used, whether or not it is specified on the drawing. Skew nailing through the connector plate at the heel invariably results in splitting of the timber in the bottom chord, simply because the nail is placed so close to the cut end of that particular member. Unless the nail is driven at an angle which would result in it emerging from the bottom of the bottom chord, it can also disturb the penetration of the teeth on the opposite side of the heel joint, thus again weakening one of the most heavily loaded joints in a trussed rafter.

British Standard 5268: Part 3: 1998, which came into effect on 15 August 1998, has unfortunately not taken the opportunity to ban this practice, although it does state that it should not be used in stainless steel plates or where the workmanship on site is not of a sufficiently high standard to ensure that the fasteners (nail plates), joints, timber

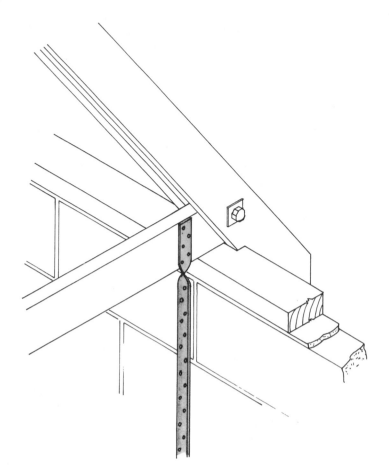

Fig. 7.11 Trussed rafter strap.

members and bearings will not be damaged by careless positioning or overdriving of the nails! In the author's opinion, it is better to design-out this possible problem by specifying the use of truss clips, the strength of which can be proven by calculation – there is no way to prove the strength of skew nailing.

GABLE ENDS, LADDERS, GABLE RESTRAINTS AND SEPARATING WALLS

Gable ends

There are three principal alternatives for the detailing of roofs at the gable ends or 'verge':

(1) To provide an overhang similar to that at the eaves complete with timber trim or barge boards and/or soffit;

(2) To simply bed the tiles onto the top of the wall, usually on a strip of durable material to support a minimal overhang to the tiles;

(3) To carry the gable wall up past the roof to form a parapet.

Detail one

The detail chosen by the architect will depend upon the building style rather than any structural considerations. Taking the first option above requires more attention to the structure because of the overhang. The provision of the supporting overhang will depend upon the roof construction used. Where a purlin is used to support the commons this can be taken through the gable walls and projected to give the overhang required, thus supporting the rafter at that point. Similarly, the ridge may be projected to support the common rafter at the top. Occasionally the wall plate itself may also be continued through the wall to provide support at the foot of the common rafter. If this is not done then some noggings will need to be built into the wall and fixed to the last common rafter on the roof proper, cantilevering out to carry the rafter of the projecting roof.

The barge board will be fitted to the last rafter, its top being level with the top of the common rafter. In the construction of some older houses the barge board was deep enough to project above the top of the rafter to a height equal to the thickness of the battens and the tiles, and then a capping timber was nailed to the top of the rafter to complete the weatherproofing. This capping is very vulnerable and of course is a significant maintenance problem, and for this reason the detail is seldom used today. Where exposed rafter feet are used it is not uncommon to leave the purlin, the ridge and any other supporting members (projecting out to carry the gable overhang) exposed to view. More frequently, however, the underside of the overhang is closed off with a soffit to match that of the main building. Where a soffit is required the soffit board can be fixed to the underside of the projected rafter, and on to battens fixed to the gable wall. The above detail is suitable with either the traditionally cut roof or the bolt and connector roof, both having substantial purlins, ridges and, if necessary, large wall plates.

The gable ladder

On trussed rafter roofs no purlin or ridge board exists, therefore alternative means of supporting the overhang are required. The 'gable ladder' is used to provide the roof support (see Fig. 7.12).

The gable ladder can be constructed on site, but is more usually provided either assembled or as a set of precut components by the trussed rafter manufacturer. It is common practice to use the same timbers in framing the gable ladder as are used in the trussed rafters themselves, thus ensuring good alignment of the roof. The gable ladder is a simple nailed assembly, itself nailed through one of the gable ladder rafters directly to the last trussed rafter on the main roof. The brickwork is then built around the 'rungs' of the ladder to fix it securely in place. Significant overhangs beyond the gable end can be achieved using this detail, but beyond about 450 mm particular care must be taken in

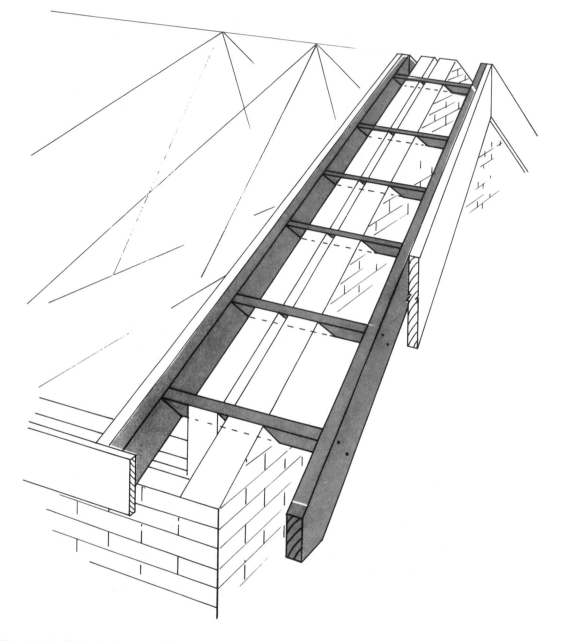

Fig. 7.12 Gable ladder – brick/block wall.

fixing the gable ladder, ensuring that it is correctly designed to carry not only the load of the roof but also the operative working on the roof. Barge boards can be fixed as previously described and, with the fascia continuing through from the main roof, the typical barge board to fascia detail can be achieved. Wind uplift on very wide verge overhangs may require the gable ladder to be strapped down to the wall.

On timber framed housing it is quite common practice to use a completely pre-fabricated verge unit, this comprising the gable ladder, prefixed soffit, and prefixed barge boards. This unit will be nailed, as illustrated in Fig. 7.13, to the last trussed rafter of the timber framed house and, supported on the timber framed gable end panel, it will cantilever over to give the desired gable end overhang. The effective cantilever should not exceed 600 mm or the distance from the timber gable panel to the first roof truss if less than 600 mm. This will ensure that the truss does not carry uplift loads from the gable ladder overhang caused by its loading of tiles and wind gust uplift. Refer to Fig. 7.13, 'x' must not exceed 'y'. Care must be taken to ensure a settlement gap between the brickwork skin and the soffit to allow the timber frame to settle independently of the brickwork without disturbing the true roof line. The gap should be filled with a compressible filler.

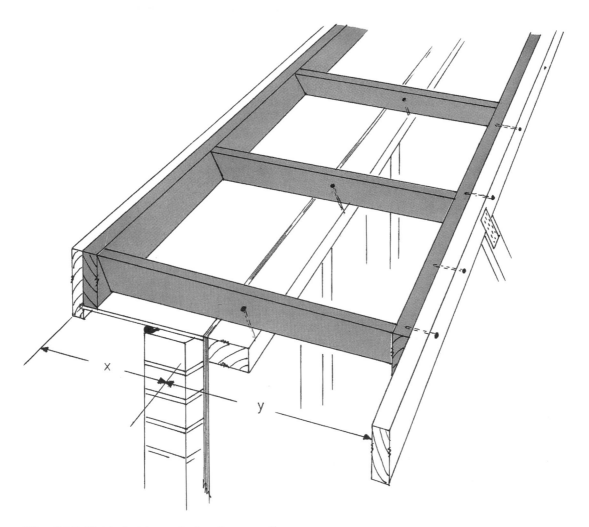

Fig. 7.13 Gable ladder – timber frame wall.

Detail two

Detail two above requires little description. There is no timber structure to the roof beyond the last common rafter on the inside of the gable end. As stated earlier, the roof tiles are simply bedded on to the brickwork gable and giving a minimal overhang, with gaps between the tiles and brickwork and the laps of the tiles simply being pointed with cement mortar.

Most of the principal tile manufacturers now have what they term to be a 'dry verge' system. These systems use plastic extruded units to form a simple but neat and weatherproof gable and trim where no barge board is required.

A further option available from the tile manufacturers is to use a special verge tile, which in visual effect folds the tiles over the end of the roof down onto the brickwork again giving a neat weatherproof and maintenance-free finish. No timber work is involved in either of the proprietary systems.

Detail three

The parapet wall presents its own problems of weathering between the wall and the roof abutment. There are no structural problems as far as the timber roof structure is concerned, the last common rafter or trussed rafter simply being placed close to the inner skin of the gable end wall. The trussed rafter in particular should not be fixed to the wall, thus allowing natural deflection to occur in the trussed rafter roof independent of the brickwork. The weathering between wall and roof is generally achieved by means of lead soakers placed between the tiles and turned up the face of the abutting wall, these being covered by stepped lead flashings fitted into the brick courses.

The detail at the abutment will be influenced by the shape and style of tile used for the main roof. Figure 7.14 shows a stepped lead cover flashing frequently used with heavily rolled tiles and pantiles.

An alternative and particularly neat detail with the flatter type of tile is to use a very small secret gutter between the roof tiles and the parapet wall. The tiles are simply stopped some 25–38 mm short of the parapet wall, with a continuous length of lead dressed underneath the tiles with a welt or water check, down on to a small supporting batten between the last rafter and the wall, and then up the face of the wall to a height approximately matching that of the tiles themselves. Dressed over the top of that is the stepped apron flashing in the normal manner.

Proprietary roof tiling systems

It is not intended to discuss here the merits of the various roofing systems offered by the tile manufacturers. This is considered outside the scope of this book, which concentrates on the timber roof structure rather than the coverings. It is useful, however, to be aware of the many roofing details produced in the various manufacturers' tiling manuals of technical literature.

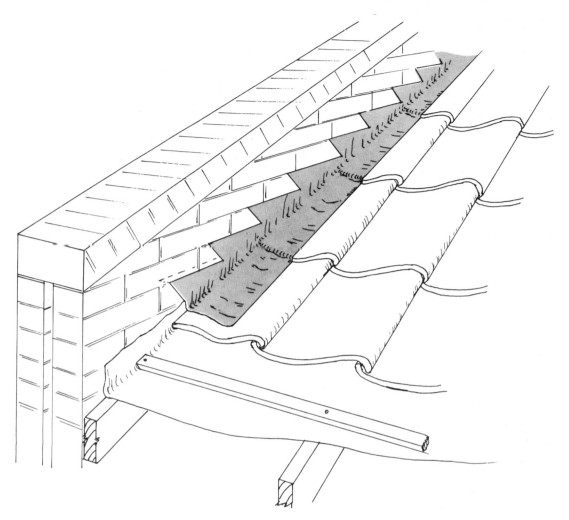

Fig. 7.14 Stepped flashing.

'Party' or separating walls

The other form of gable occurring in roof structures is that unseen gable constructed at the division between dwellings in pairs or terraces of houses. The wall in that position must be continuous to the underside of the tiles to ensure adequate fire break, and generally it is not acceptable to have timber members built into this wall, unless they are adequately separated between the houses with a material which will impart one hour fire resistance. Good practice would therefore dictate that timber in general is not built in to separating walls.

A problem which occurred in the early days of truss and trussed rafter use in this country was that of 'hogging' over these separating walls between buildings; this is illustrated in Fig. 7.15. To overcome this problem, the party wall brickwork or blockwork must be kept down below the top of the rafter line by some 25 mm. BS 5268: Part

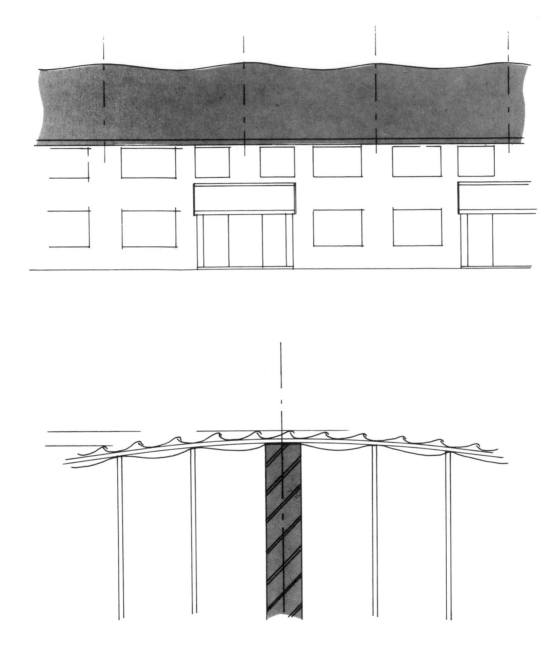

Fig. 7.15 'Hogging' at compartment wall.

3 limits deflection to 12 mm for roofs up to 12 m span. Consequently at maximum deflection of the trussed rafter roof there is still a 12 mm gap between the top of the rafter and the top of the brickwork. This gap must be filled with compressible yet non-combustible material (usually mineral wool), the tile battens themselves being the only timber item to pass from one building to the other. The detail described avoids the problem of hogging and its unsightly effect on the roofs of terraced houses. Figure 7.16 illustrates the correct detail.

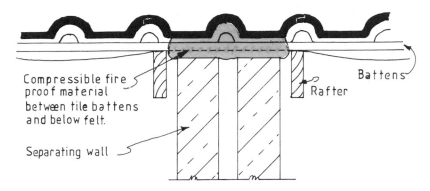

Fig. 7.16 Fire stop at compartment wall.

Gable wall restraint

The question of pressure on one end of the gable end of a house and suction on the other end was discussed in Chapter 5. Figure 7.17 illustrates this effect.

Above the wall plate the gable end is of course free standing brickwork and on steep pitched roofs this area of gable brickwork can be quite large, and the resulting

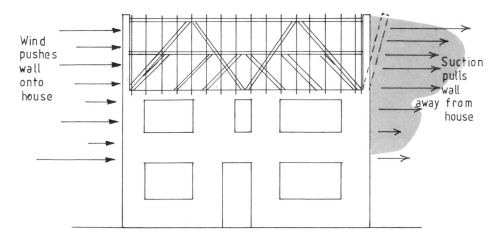

Fig. 7.17 Wind loads on gables.

wind loads quite significant. The Building Regulations require these external walls to be restrained and BS 8103/1 gives recommendations for the horizontal lateral connection between masonry walls and the roof structure. NHBC, in its section 7.2 referred to earlier, covers this aspect under 'strapping' clause 7.2, S3. With the above ties and bracing as set out in BS 5268: Part 3, annex A, described in Chapter 5 of this book, the roof will be resistant to all lateral loads imposed upon it, provided the limiting criteria set out in the standards are not exceeded. Similar bracing and restraint will apply to all roofs, however constructed, and this must not be forgotten simply because the details which are now generally available invariably show only trussed rafter roofs. With a traditional roof the restraint of course can be provided more simply by ensuring good fixity for purlins, ridge, and other longitudinal members into the gable ends.

Restraint is generally provided in trussed rafter roofs by fixing galvanised steel straps across at least three trussed rafters into the gable end. These contain the suction loads holding the brickwork into the roof, whilst blocking pieces fixed between the wall and the first trussed rafter and between the first trussed rafter and the second trussed rafter guard against the gable being blown into the roof. The ties should be of 30 mm × 5 mm minimum in cross-section and are generally prepunched with nail holes to facilitate ease of fixing on site. Figure 7.18 illustrates a typical set of gable restraints.

It can be seen that these straps occur at both ceiling and rafter on gable ends, but on the ceiling they tie only on separating walls where they pass through them to form a continuous link between adjoining roof structures. The ties should be fixed with 3.35 mm diameter, 50 mm long galvanised round wire nails.

In extremely exposed situations, care must be taken to check that the parameters laid out in the British Standard are not exceeded; if so then a separate strapping design will be required. The standard spacing for straps is 2 m on both rafter and ceiling tie, but this may have to be reduced to provide more straps if additional restraint is required. The straps may also need to be longer, being attached to four adjacent trussed rafters rather than the standard three.

WATER TANK PLATFORMS

Another load frequently supported by the roof structure is the cold water storage tank and also possibly the heating system expansion tank. These loads must be taken into consideration when designing and constructing the roof. A check should be made with the local water authority for their precise requirements on the tank size required. The capacity of course dictates the weight of water to be supported, whilst the size may control the location of internal roof members. The capacity will usually be 230 or 330 litres. The NHBC have specific requirements for the access to and around water tanks in roof spaces and these are clearly stated in the *Registered House Builder's Handbook*, Vol. 2, Chapter 7.2, clause D14. Briefly, access boarding must be provided from the loft access to the tank stand, and at least one square metre of boarding must be provided around each tank.

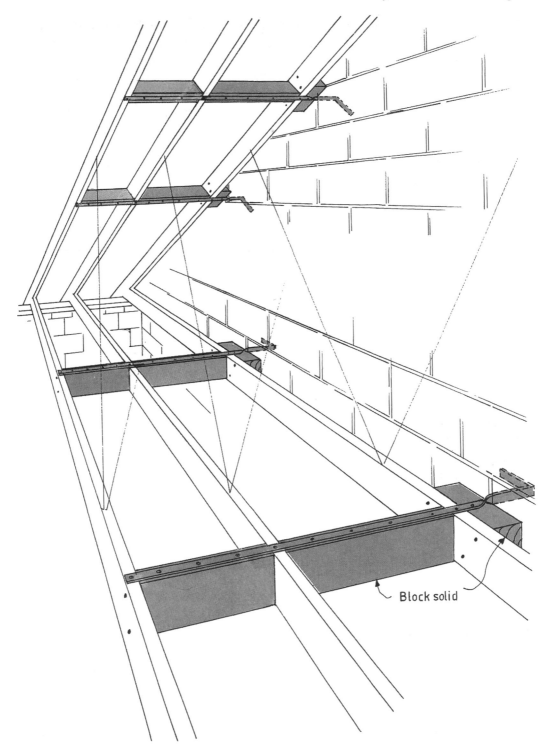

Fig. 7.18 Gable restraint straps.

In traditionally constructed roofs tanks have usually been, in the past, supported directly from walls below independently of the roof, or on a stand bolted to the gable brickwork. If supported directly from the walls below a detail similar to that shown in Fig. 7.19 could be used, but beams B must be designed to carry the load from the supporting walls which would themselves be replaced by beams A. The gable wall supported stand must of course have a specifically designed, probably steel bracket system which itself must be very securely fixed to the wall structure.

If the roof is to take the weight of the tanks and their supporting platform, the load will be placed invariably on the ceiling joists which are, of course, supported by binders and hangers from the purlins. All of these members will have to carry the additional weight and, moreover, careful attention must be paid to the fixings between them. It will be found necessary to use bolted connections rather than simple nails to transmit the loads from one member to another. Again a tank platform similar to that indicated in Fig. 7.19 could be used with beam A being supported by hangers from the purlins above.

The bolt and connector truss roof

For the bolt and connector truss roof a separate design should be prepared for the tank platforms with as much of the loads as practical being carried by the principal truss itself. The TRADA designs do not allow for tank loads and therefore any tanks carried in roofs constructed using the standard design sheets must be supported from walls below. One way of overcoming this problem is to close the centres of the principal trusses in the area of the tank thus effectively strengthening the roof in that area, enabling the structure to carry the tanks independently of walls below. Design advice must be obtained from TRADA. In such instances the design shown in Fig. 7.19 can again be used.

The trussed rafter roof

In trussed rafter roofs, BS 5268: Part 3: 1998 requires the building designer to advise the trussed rafter designer of 'the size and position of all water tanks and other ancillary equipment or loads to be supported on the trussed rafters'. This is clause 11.1g. We can assume then that the trussed rafters will have been designed to carry water tanks and provided they are within prescribed limits and location within the truss, a standard tank support platform can be used.

Figure 7.19 shows a typical tank stand constructed to the guide-lines laid down by BS 5268: Part 3, but for precise details of timber sizes reference should be made directly to the British Standard, or to the TRA technical literature, or to any of the trussed rafter plate manufacturers' technical brochures, or again to NHBC pitched roof clause 7.2, appendix F. Depending on the tank size, it must be supported over three or four trussed rafters, always with bearers A as close as possible to the node points of the truss. A method of lowering the tank stand where restricted headroom is a problem is shown in the TRA *Technical Handbook*.

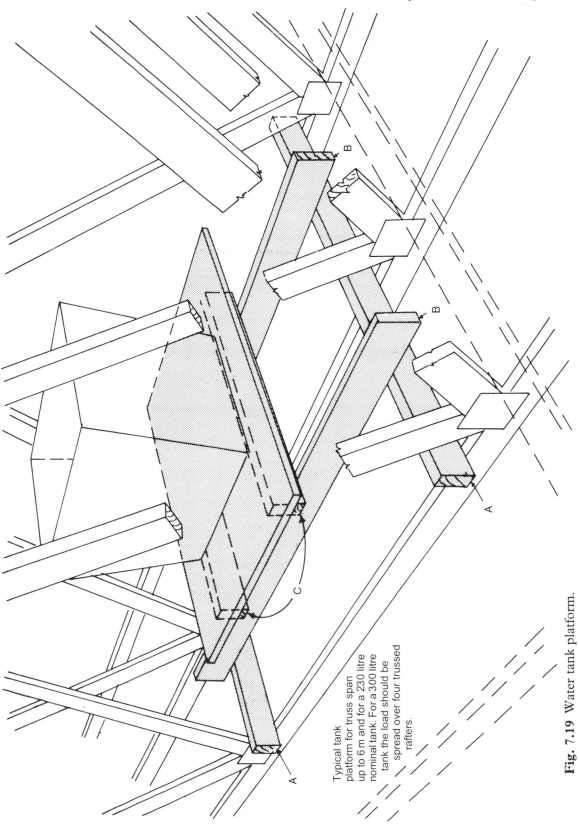

Typical tank
platform for truss span
up to 6 m and for a 230 litre
nominal tank. For a 300 litre
tank the load should be
spread over four trussed
rafters.

Fig. 7.19 Water tank platform.

The question of restricted headroom must be considered when deciding upon the location of the tanks within the roof. The tank will normally be supported about the centre line of the roof truss, this generally being the largest clear void within a trussed rafter roof. Low pitched roofs bring their own problems of restricted headroom, likewise the numerous timber members in a hip end roof generally prevent tanks being placed within the hip area. Not only must access to the tanks be considered, but also reasonable space must be left around the tank for initial installation and maintenance thereafter. Do not forget the additional thickness of tank insulation. NHBC NHB standards pitched roof clause 7.2/S14 advises on this subject.

General considerations

Where showers are installed in rooms below, it is often required to raise the level of the tank as high as practical to afford head of water for the shower. The tank platform indicated in Fig. 7.19 can still be used, but modified by building a rigid ply clad box-like structure around the above supports C. On the top of this box the normal water tank platform itself may be securely fixed. *Do not nail supports B to the sides of the trussed rafter webs*, they are not designed to carry such loads and the results could eventually be disastrous.

In refurbishment work, where the plumbing system is often renewed, the designer and/or installer must make sure that adequate ventilation is provided for the roof space to prevent condensation occurring on pipes to, and the tank itself. This condensation can be a cause of seriously wetting the tank platform structure, causing degradation of both platform and supporting bearers. If chipboard is used for the platform then it should be of the moisture resisting grade.

VENTILATION OF ROOF VOIDS

The ever increasing need to conserve the energy used for home heating has led to higher insulation standards in recent years, and further improvements in these standards can be anticipated in years to come. The insulation is usually placed between ceiling joists, resulting in what is known as a cold roof space, as against a warm roof space, which would be the case with insulation placed within the rafters. The following discussion assumes a cold roof space situation.

Moisture vapour rising from the home below passes through the ceiling and insulation into the cold atmosphere of the roof and will, under certain conditions, condense into water on the coldest elements within the roof space. These cold elements are invariably the metal fixings used to construct the roof and of course, in the case of a trussed rafter roof, the nail plates used for the trussed rafter assembly itself. Clearly this wetting can lead to deterioration of the metal fixings and, if continued for long periods, to the deterioration of the timbers if they are not fully preservative treated.

The solution

Two possible solutions exist, at least in theory. Firstly, prevent the moisture-laden air entering the roof space or, secondly, ventilate the moisture-laden air before it can condense and cause any harm. Examining the first option, this at first seems the most simple method, and in theory this can be achieved by placing a vapour barrier of polythene or similar vapour proof sheeting immediately above the ceiling finish, yet beneath the insulation. This would contain the moisture-laden air within the building below. In practice, however, there are numerous small holes through this vapour barrier in the form of electrical services, hot and cold water services and soil and vent pipes. There is also the problem of effectively sealing joints between the sheets of polythene used. The major drawback to this option in the writer's opinion is that a vapour barrier immediately above the ceiling finish effectively traps some of the vapour, and can lead to unsightly mould growth which can be extremely difficult to eradicate once it has appeared. It is virtually impossible to contain moisture within rooms such as bathrooms, shower rooms and kitchens and therefore this first option of a vapour barrier is not a practical one.

We are therefore left with the option of ventilating the roof void. The British Standard 5268: Part 3 no longer sets out minimum requirements as this is considered the responsibility of the 'building' designer as distinct from the roof structure designer. BS 5250 gives guidance on prevention of condensation in roofs. NHBC cover the subject in clause 7.2/S11(a), *Ventilation*. The Building Regulations themselves in approved document F clause f2, *Condensation*, also require the designer to take account of the possibility of condensation within the roof space. The TRA technical bulletin gives details of how this ventilation should be provided in conjunction with trussed rafter roofs.

Satisfactory ventilation will be provided by designing a minimum gap of 25 mm along at least two opposite sides of a roof where the pitch does not exceed 15°, or 10 mm for roof pitches above 15°. Furthermore, when a monopitch roof is being considered or a duopitch roof in excess of 20°, or 10 m span, consideration should be given to providing a further 5 mm of continuous ventilation at the ridge. Care must also be taken to identify condensation traps such as those below dormer and roof-window cills, providing some 5 mm of ventilation at those points. Do not forget to introduce ventilation above the dormer roof or roof window head, remembering also that the dormer roof itself should be provided with ventilation. There are many ways of effectively providing this ventilation, mostly via the soffit in the form of slots or holes covered with an insect-proof gauze. Having introduced the airflow into the soffit void, care must be taken to prevent insulation blocking the space between the rafters, thus preventing this flow of air into and through the roof space. In addition, therefore, some method of controlling the insulation must be introduced.

Proprietary systems

The two leading roof tile manufacturers in this country, namely Marley and Redland, both have their own systems for providing ventilation at eaves level and at ridge roof spaces. Both systems use lightweight plastic type mouldings to provide the ventilation, insulation control and the necessary insect screening. Both companies produce systems where the ventilation is provided on top of the fascia and not through the soffit by, in Redland's case, a single plastic moulding, and in Marley's case a set of mouldings which at one and the same time provide ventilation, support the roofing underlay felt to prevent ponding and control the insulation. Both systems have a ridge ventilation method which is 'dry fixed' meaning that it does not depend on traditional cement mortar for secure bedding.

Other companies manufacture individual ventilators and insulation controllers, the latter usually designed for use with trussed rafters at the standard 600 mm centres. All provide components complying to the necessary requirements, but careful specification of the individual item is necessary to ensure that the correct air gap for the pitch being used is achieved. Figure 7.20 shows the essential features of eaves ventilation and insulation control.

The attic roof

The attic roof of course has three roof voids, those at either side at low level and the triangular roof void above the attic room. Airflow must again be introduced at low level, i.e. at the eaves, and allowed to flow unobstructed from the lower roof void up between the rafters to the upper roof void, where exhausting from the ridge is particularly important because of the usually steep pitch nature of the attic roof.

The area of rafter which forms the sloping ceiling of the room must be carefully considered. When attic trussed rafters are used then this rafter may well be 200 mm deep which, when 100 mm of insulation is used between the rafters, should allow adequate ventilation space above the insulation from one roof void to the next. However, unless the designer can be absolutely certain that the insulation will not slide down the sloping ceiling, some controls may need to be introduced to ensure that adequate ventilation space of 50 mm between each rafter is maintained.

On traditionally constructed attic roofs where the timbers may be smaller, or in the type of construction illustrated in Figs 5.19 and 6.15, the timber used for the rafters will be relatively small. To ensure that adequate insulation thickness and air space are maintained, it may be necessary to specify a minimum rafter depth of at least 175 mm, thus giving room for 100 mm of insulation, an insulation controller and a 50 mm air space. In the next section, this subject of insulation control is again highlighted in the discussion on sheet material used for stability bracing in attic roof construction.

BRACING

Any roof constructed of simple timber members or manufactured components is about as stable as a row of dominoes standing on their ends. Similarly, any pressure on the end of these components will result in a 'domino effect', each toppling on to the other until possibly total collapse occurs. The roof clearly has to be braced against this effect and it is not adequate to consider the brick and block gable ends satisfactory for this purpose.

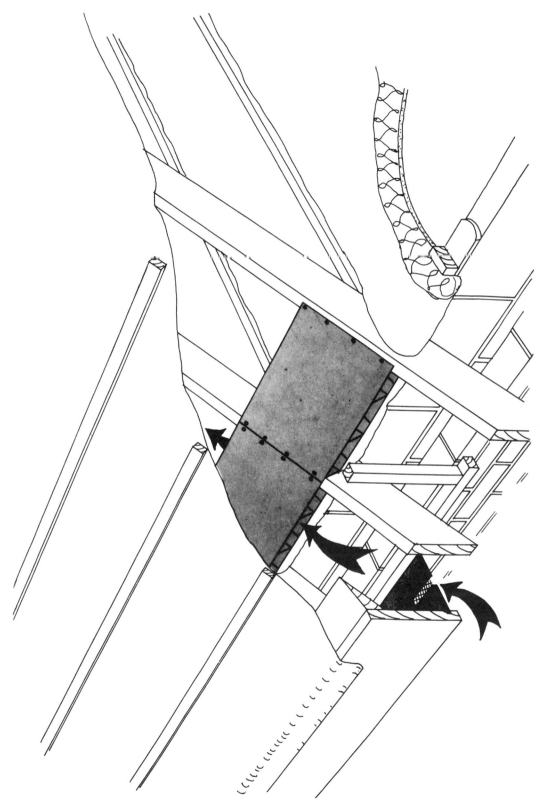

Fig. 7.20 Roof ventilation.

The section on gable end restraint should have emphasised this point adequately.

There are three main types of bracing to be considered:

(1) Temporary bracing required to stabilise the roof whilst it is actually being worked on and in the construction stage, leaving it secure during breaks in construction. Many partly built roofs have been damaged by sudden storms when left in an unstable condition between work periods.
(2) Stability bracing, required to brace the individual timbers in position in the finished roof and to keep the roof itself upright.
(3) Wind bracing, required to enable the roof structure to resist loads imposed upon it by wind acting on gable ends, or in some instances on the supporting walls below.

These three types of bracing apply to all forms of roof construction (refer to Fig. 7.17).

A fourth type of brace, the 'Web Compression Brace', may be required by the truss designer. See page 78 for additional detail; item J.

The traditional cut roof

Refer now to Fig. 7.21. On traditionally constructed roofs little temporary bracing is required if the purlins are well fixed at their support points. This means of course ensuring that supporting brickwork has not only been constructed but that the cement mortar is adequately cured, otherwise damage to the wall could occur from the movements of constructing the roof above it. Although the illustration shows the trussed rafter roof detailed in Chapter 5, the essentials of bracing of course remain the same in any roof form. No specific guidance exists for the bracing of traditional roofs, but the recommendations laid down in BS 5268: Part 3 for trussed rafter roofs could well be followed with few modifications.

It is advisable to fit the diagonals B, the binders G being replaced by the ridge board and the ceiling joist binders themselves. Stability bracing can be achieved by adding diagonal brace F with brace J being fitted from the binder diagonally across the hangers to the purlin. Because the hangers may well be spaced further apart than with a trussed rafter system this member should be more substantial, probably 50 mm × 100 mm in section. If the purlin strut is used as illustrated in Fig. 3.5 this additional brace will not be required. Brace H in Fig. 7.21 is of course the purlin itself in a traditional roof. Finally the ceiling joist diagonal brace K should be fitted as indicated.

The bolt and connector truss

The principal trusses are significantly heavy and great care must be taken to temporarily stabilise the first pair of these trusses fixed in position. Again referring to Fig. 7.21, braces B should be installed as early as possible but should be of 50 mm × 100 mm cross-section and well nailed to both the trusses and the wall plate. Furthermore, a prop should be fitted under the ridge collar down to the first floor joist system (or floor if a

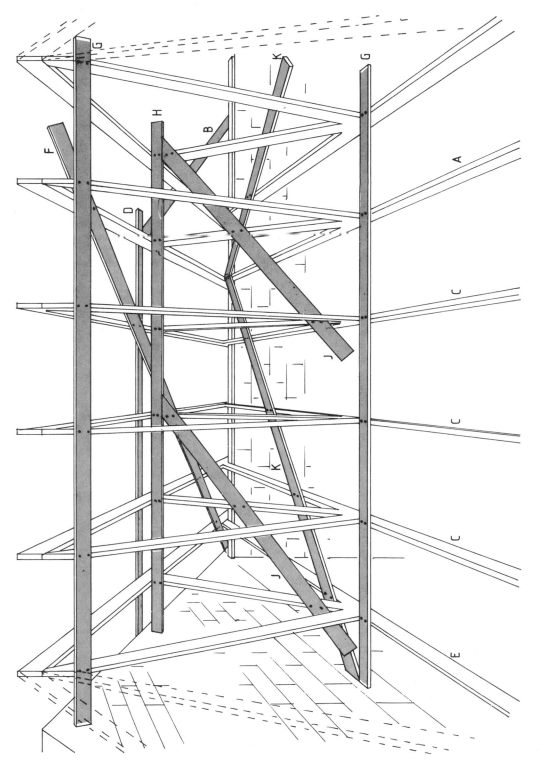

Fig. 7.21 Stability bracing.

single-storey building) below. Because of its length this prop should be a scaffold tube fitted with a suitable clip to lock it on to the ridge collar, and restrained at the bottom by timbers laid flat and nailed to the floor joists, or fitted tight against a cross wall. Apart from this precautionary measure resulting from the weight of the trusses, the other items of bracing can be considered as for the traditional roof described above.

The trussed rafter roof

Bracing for the trussed rafter roof has been fully described in Chapter 5 because of its inclusion in BS 5268: Part 3 which was analysed in that section, and because of the trussed rafter's dependence upon the bracing system for its overall structural integrity.

Hip roofs

Hip roofs are essentially well braced structures by virtue of the hip board diagonal braces inherent in the design.

Again referring to Fig. 7.21, binders G and H should be fitted to the section between the hip ends and as far into the hip ends as possible. Diagonal K should also be fitted and, unless the distance between the hip peaks is very long, diagonal F should not be required.

Valley intersecting roofs

Each roof section should be treated individually with the guidance set out above. The valley boards themselves add extra diagonal stiffening in that area. The discontinuity of tile battening under the abutting roof should be replaced by timber binders placed on top of the common rafters as described in trussed rafter valley roofs in Chapter 6.

Attic roofs

Attic roofs, because of their great height and the inherent large void forming the rooms, present their own particular problem with regard to bracing. The trussed rafter attic has been dealt with in some detail in Chapter 5. It is unlikely that an attic will be constructed using bolt and connector principal trusses. The notes below therefore apply to traditionally constructed attics and also generally to trussed rafter attic roofs, although in the latter case specific designs for bracing should be obtained from the truss designer.

Figure 7.22 should be referred to. For stability reasons, the gable walls will probably not be constructed above the upper purlin level at the time roof construction starts, and it is assumed in this description that the long rafters will not be available in one length, and have been split into the upper and lower rafter sections, namely A and J respectively. The description also assumes that the purlins and floor joists are fixed in position.

Construction should commence by fixing six of the lower rafters A from wall plate to upper purlin on both sides of the roof. Temporary diagonals B should now be fixed in place on both sides of the roof, securely fixed to each rafter passed. Collars or ceiling ties

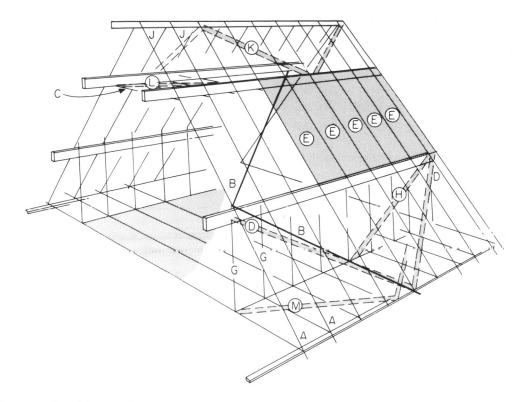

Fig. 7.22 Stability bracing – attic roof.

C should now be fitted to stiffen the upper purlins. Next complete the commons to the lower part of the roof and the collars. Diagonal braces D should next be fixed to the underside of rafters A. Next fix uprights G between purlin and floor joist to form the walls of the room. Care should be taken to set these uprights or studs at centres appropriate to the wall covering board to be used. Diagonal H should now be fixed across the end five or six uprights, nailing this to the void side of the stud, and nailed to each stud it passes.

Having stabilised each lower section of the roof, attention must now be turned to the space between the purlins, to ensure a solid structure before completing the roof to the ridge. To avoid a diagonal between the purlins on the underside of rafters A a sheet material E, which is suggested to be 9 mm or 12 mm regular sheathing ply, should be tightly fitted between the common rafters A for approximately five rafter spaces away from the gable end. This ply should be nailed to battens fixed to the side of rafters A at their top edge. The plywood, being fixed to the underside of these battens, will then act as insulation controller in this sloping part of the ceiling to the roof below, yet maintain an air space above it and below the felt underlay to the tiles. The batten depth therefore must be a minimum of 50 mm. This ply diaphragm, provided it is adequately nailed, will impart great stiffness to this otherwise unbraced part of the roof, the ply panels being fitted on both sides of the slope and at both gable end situations. The upper section of these panels may have to be left off until the upper rafters J are nailed in position on the purlin and lapped to the top of rafters A.

Finally rafters J and their ridge can be fitted, the rafters being birdsmouthed over the purlin in the usual manner and well lapped and nailed to the top of rafters A. Racking sheathing E can now be completed. Diagonals K and L can now be fitted to the underside of the rafters and to the top of the ceiling collars C.

All timbers mentioned above for bracing can be of 22 mm × 97 mm section, with the specification for nailing and lapping, if necessary, as set out in Chapter 5.

The final brace for an attic roof is of course the floor boarding or sheeting applied on top of the floor joists between the uprights G. This floor diaphragm is subjected to more loading by the structure above it than with a conventional two-storey building. It is essential therefore that the floor joists are solidly bridged between them, or fitted with herringbone strutting, and that supports are provided to carry the floor boarding, particularly at the junction with the wall G. Correct nailing of any boarding used for the floor must be used; if chipboard is used this would mean annular ring shanked nails and gluing of the tongued and grooved joint between the boards.

The effect of openings on bracing

All attic roofs will have some openings in the sloping part of the ceiling, either in the form of roof lights or dormers. The effect of these openings on the bracing will naturally depend upon the frequency and size of opening in either or both roof's slopes in relation to the overall length of the building. Clearly the provision of adequate bracing in the form of the panels E must be considered and if this cannot be provided immediately adjacent to a gable wall, then continuity between the panels near the wall and those repositioned because of an opening must be maintained to allow the two separate areas of bracing to act as one.

Stairwell openings within the main body of the building are unlikely to cause any significant decrease in the effectiveness of the floor diaphragm. Those staircases constructed parallel with the floor joists and immediately adjacent to a gable wall, however, may present stability problems for that gable wall which will invariably be supporting one of the purlins carrying the roof. A means of providing an alternative to brace H must be found and additional attention may be needed to gable end restraint in such instances, above and beyond that dealt with in the section on gable wall restraint in this chapter.

EAVES DETAILS

The eaves of a building vary greatly in design throughout the UK, and to some extent are an architectural detail rather than a structural requirement. The function of the eaves is of course to close off the ends of the rafters and, where a generous overhang is provided, to protect the building to a certain degree below. The traditional large overhang associated with most thatched roofs provided excellent protection to the heads of doors and windows below.

The functions to be considered in the design of eaves are therefore as follows:

(1) To effectively close off the space between the rafter feet;
(2) To provide a means of ventilation for the roof;
(3) To provide protection for the building below if required;
(4) To provide support for the rainwater drainage system;
(5) To provide support for the tile underfelt;
(6) To provide support for the soffit if required.

One of the most important features mentioned above is the support of the tile underfelt. Figure 7.23a shows the problems of underfelting being unsupported, being allowed to sag without support between the rafters and thus allowing ponding, with the resulting degradation of both fascia and soffit and possibly the top of the wall structure itself. Adequate support must be given at the bottom of the roof slope for the felt to avoid this ponding, this being achieved in the form of a thin sheet material applied to the top of the rafter feet or sprockets if provided, or in the form of a continuous triangular fillet fixed to the top of the rafter feet. This detail (Fig. 7.23b) allows any water which may have penetrated through the tile to run down the roof slope into the gutter in the normal manner.

The next important aspect is to detail the eaves allowing adequate ventilation, and simple methods to achieve this are indicated in Figs. 7.24a–e.

Figure 7.24a shows a detail with no overhang, care being taken not to fix the fascia tight to the wall, although with a ventilation system shown in Fig. 7.24d the fascia could be fitted directly to the wall if necessary. Figure 7.24b shows a typical overhang with fascia and soffit, this particular detail indicating a timber framed structure, care having been taken in this instance to show a gap between the soffit and the top of the brickwork to allow for the differential movement between it and the timber structure. Figure 7.24c shows a corbel eaves detail with no soffit and with the junction between rafter and ceiling tie taken beyond the outside of the wall. This particular detail would impose certain structural problems for all types of roof construction, and may require a blocked heel or additional top chord, should trussed rafters be specified. This particular detail indicates a loose fill insulation with a timber board controller to prevent the insulation spilling through into the cavity, or across the cavity blocking the ventilation space.

Figure 7.24d shows one of the proprietary combined ventilators and insulation controllers fitted on top of the fascia. This particular detail also indicates a sloping soffit fitted directly to the underside of the rafters. Figure 7.24e indicates exposed rafter feet with ventilation provided by slots between the infills between rafters.

The above illustrations show only a few of the many variations on design imparting individuality to any building. The only two details likely to give any structural problems are those indicated in Fig. 7.24c because of its cantilevering effect for the truss, and any of the details where the rafter overhang is excessively long. In general this would mean beyond 700 or 800 mm, depending on the rafter depth. The use of the triangular sprocket piece on top of the rafter foot will not aid its strength in this respect, unless of course it is carried up the rafter well beyond the wall plate position.

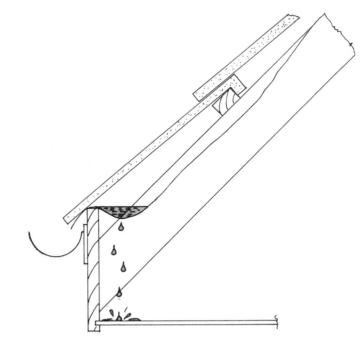

Fig. 7.23a Incorrect eaves detail.

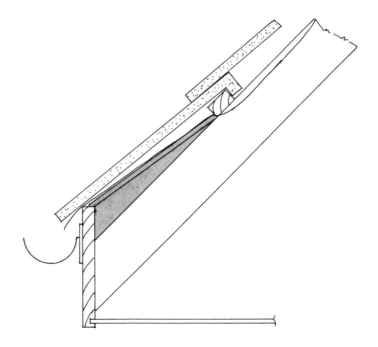

Fig. 7.23b Correct eaves detail.

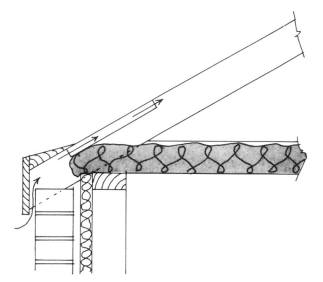

Fig. 7.24a Eaves ventilation – no soffit.

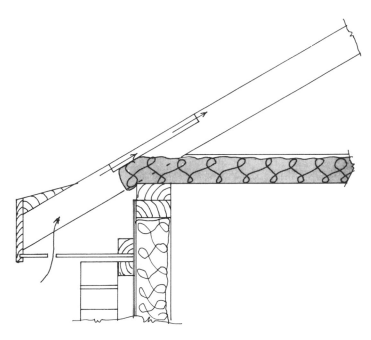

Fig. 7.24b Eaves ventilation – with soffit.

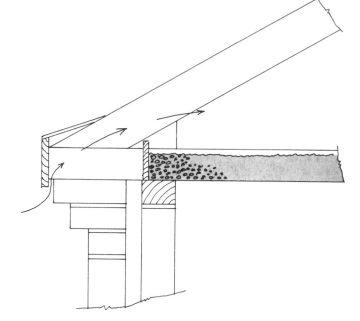

Fig. 7.24c Eaves ventilation –
corbel soffit.

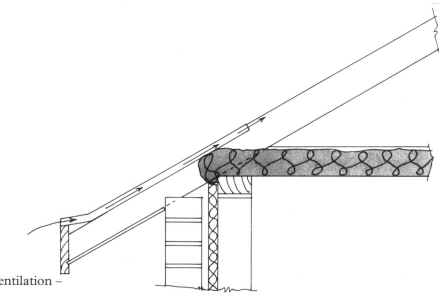

Fig. 7.24d Eaves ventilation –
over fascia.

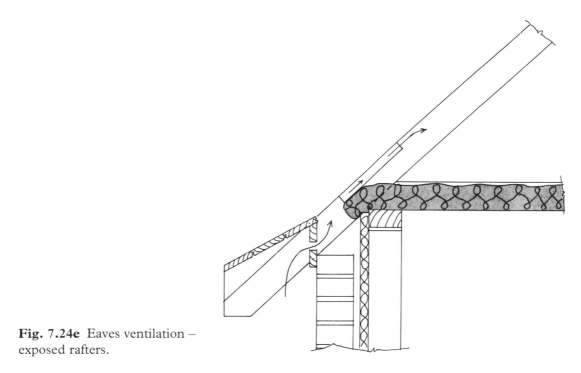

Fig. 7.24e Eaves ventilation –
exposed rafters.

Fascias and barge boards should always be preservative treated in accordance with building regulations and NHBC requirements and should be given one coat of either paint or stain prior to fixing.

The soffit boards need not be preservative treated for they are generally not exposed to the weather, although in the writer's opinion it is desirable to do so if softwood tongued and grooved boarding is used in an exposed eaves detail, such as Fig. 7.24e. It is normal to support the soffit at the fascia by fitting it into a groove in the back of the fascia, and on light timber softwood framings on the wall side of the building. Figure 7.20 illustrates a well framed soffit support system.

Exposed rafter feet

Exposed rafter feet tend to be a fashion feature, but are very common in rural areas where new buildings need to blend architecturally with those older dwellings surrounding them.

Where trussed rafters are used in conjunction with a detail such as that indicated in Fig. 7.24e, it must be borne in mind that the timber will be planed not sawn, as will have been traditionally the case, that it is likely to be stamped with bright red or other colour stress grade marks, and that for economy reasons the trusses (and therefore their exposed rafter feet) are likely to be spaced at 600 mm centres compared to the more normal 400 mm centres of the traditional buildings.

Assuming that the centres of the trussed rafters are acceptable, the grade marks may be overcome by the use of a dark stain or indeed any paint system, but light stains will not be adequate to conceal the marks. To overcome the problem of centres and the fact that the raft feet are planed, the detail indicated in Fig. 7.25 could be used to give a more authentic eaves detail. This allows the economy of the trussed rafter to be placed at 600 mm centres. It also allows the architect to choose the precise centres at which he would like to place the rafter feet and allows him to use sawn timbers of possibly a heavier section than the timbers used for the trussed rafter construction. This particular detail

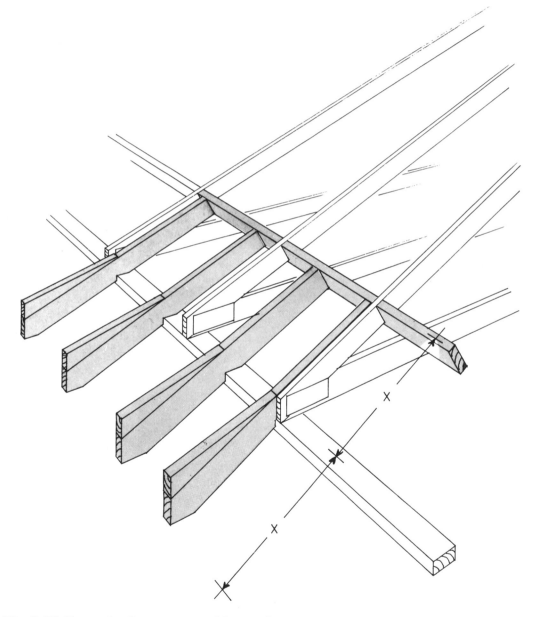

Fig. 7.25 Exposed rafters at eaves with sprocket.

also solves the problem of irregular trussed rafter spacings which invariably occur around hip end roofs. It is suggested that the dimension X should be equal on either side of the wall plate position. By birdsmouthing the supplementary rafter feet over the wall plate, a deeper rafter than that used for the truss rafter itself may be used if desired.

TRIMMING SMALL OPENINGS

The openings dealt with here are those required for small flues and loft access hatches, as well as the smaller openings for roof windows. Large openings have been detailed in the relevant preceding chapters.

In designing the roof, careful consideration should be given to the location of any flue or chimney passing through the roof void, and its likely effect on the structure. Similarly the loft hatch should be considered, particularly in relation to the spacing of trussed rafters. Roof windows for fenestration reasons may have to take priority over the structure, but again with consideration, the readily available 550 mm wide roof windows should be used if possible to fit neatly between trussed rafters, thus achieving ultimate economy in the structure by avoiding trimmed openings.

Where brick flues pass through roofs, the Building Regulations stipulate that 40 mm minimum must be allowed between the structure and the brickwork face. Floor boards, skirtings, tile battens, etc., may of course touch the brickwork but not the actual structure itself. Please refer to the relevant Building Regulation clause for precise details with regard to different classes of appliance and flue sizes.

The Building Regulations do not stipulate minimum size for loft hatches, but the NHBC do in their *Registered House Builder's Handbook*, Vol. 2, Chapter 7.2, clause D14. This requires a minimum size of 520 mm clear in any direction. The clause also states that the opening should not be directly over stairs or in any other hazardous location, such as close to eaves where headroom in the roof space is limited.

Traditional roofs

In traditional or bolt and connector roofs, small openings may be created in either the rafters or ceiling planes by simply doubling the rafters on either side and trimming, as indicated in Fig. 7.26. This would be considered suitable for openings up to 1.2 m wide, and for a length the maximum distance between two purlins or purlins and plate. Larger openings must be dealt with as set out in Chapter 3.

Trussed rafter roofs

Trussed rafters of course are designed to work at 600 mm centres (or some other specified dimension), and these spacings should not be increased without adjusting either the design of the roof truss itself, or the spacing on either side of the opening created. On no account should a trussed rafter be cut.

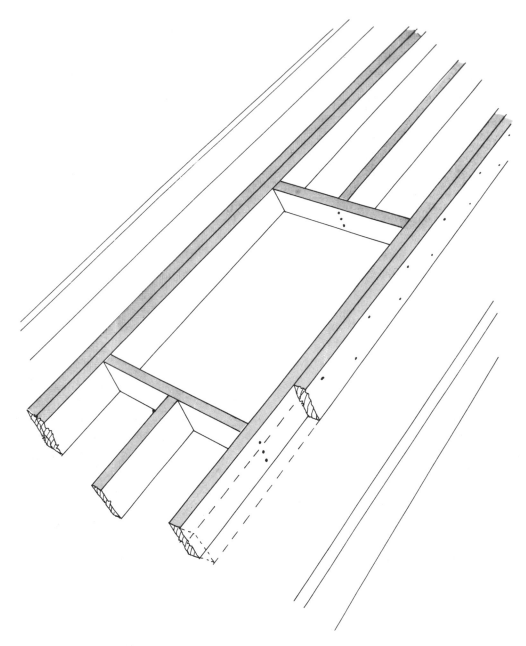

Fig. 7.26 Roof light trimming.

British Standard 5268: Part 3, pages 23–25, gives details of openings for chimneys and hatches. Section 7.6 of the standard sets out details for the maximum spacings between trimming trussed rafters. Using the British Standard's lettering, the standard trussed spacing equals a, the spacing between trimming trusses and the adjacent truss

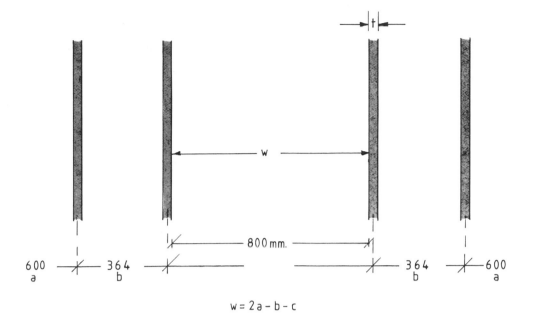

$$w = 2a - b - c$$

Fig. 7.27 Trimming for small openings – trussed rafters.

equals *b*, with the distance between the centres of the trimming trusses being *c*. This gives a formula of $c = 2a - b$. The nominal opening *c* is not that which the designer would need to know, so let the *actual* opening between the trimming trussed rafters be *w*. Assuming then a truss thickness of *t*, the actual opening width between trussed rafters becomes $2a - b - t$. To find the maximum opening width permissible for a truss spacing *a* = 600 mm, with a truss thickness *t* = 36 mm, we have $2 \times 600 - 36 - 36 = 1128$ mm. The dimension *b* must be equal to the truss thickness *t*, because to give the widest opening the trimming truss must be immediately adjacent to the last standard truss.

Let us now assume that we need an actual opening of 800 mm, and *w* = 800, *a* = 600 and *t* = 36. Substituting in the formula above we have 800 mm = $2 \times 600 - b - 36$, $b = 1200 - 800 - 36$, *b* = 364 mm. The setting out of this truss would then be as shown in Fig. 7.27.

INFILL

Infilling between the trussed rafters should be carried out with timbers of similar size to those for the trussed rafter members themselves supported on the wall plate, a purlin which will effectively be a heavier piece of timber nailed to the webs in the position of brace H (Fig. 7.21), and to a timber member nailed between the trimming trusses on either side of the opening concerned. All of this assembly is carried out using galvanised nails, but for the 'trimmers' supporting infill between trusses it is strongly recommended that these are fixed to the trimming trusses with small metal framing anchors. Figure 7.28 illustrates a typical chimney infill situation.

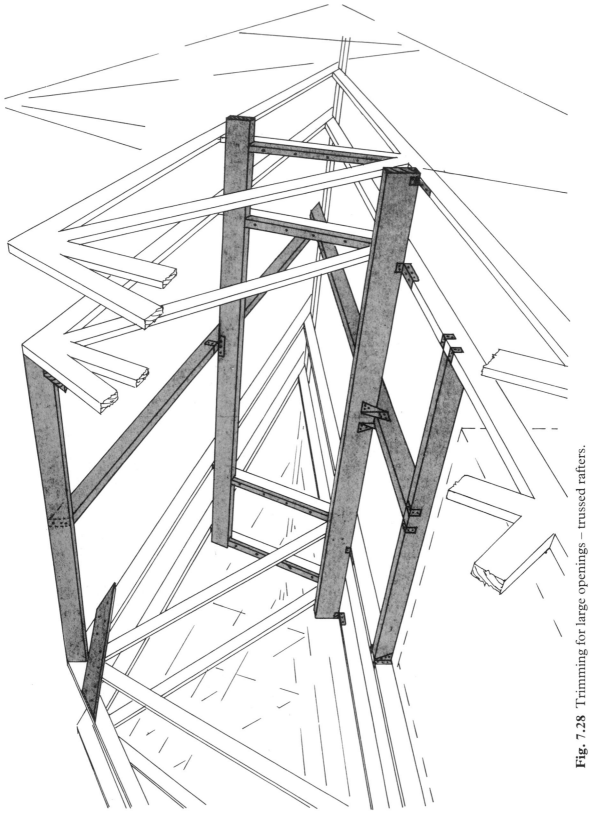

Fig. 7.28 Trimming for large openings – trussed rafters.

CHAPTER 8

Loft Conversions

Chapters 3–7 of this book have dealt with the construction of new roofs on dwellings. There is, however, a growing interest in converting loft spaces to additional accommodation, this often being the most cost effective 'extension' to the home. This and the following chapters give guidance on such conversions from initial survey to construction.

DEVELOPMENT OF THE LOFT

Utilising the roof space for living accommodation is not new. Although the very earliest simple coupled roofs would not have been strong enough to support the floor beneath, by the fifteenth century the attic had begun to appear. Very early dwellings had no chimneys as we know them: the additional height afforded by an open lofted room allowed smoke to rise above the living space to vent through a hole at the ridge.

The combined factors of bricks becoming more available and affordable and the demand for larger roof spans led to the construction of a centre support which often took the form of the chimney stack. The large chimney stack was frequently the first part of the house construction undertaken, this central structural unit later being developed to carry both floor and roof load.

As houses became larger there tended to be more rooms on the ground floor, and, with chimneys removing the need for lofty ceilings because of better smoke removal, it seemed logical to utilise this roof space by constructing a simple floor. This floor was often supported on the tie beam of the principal truss. These floors were often known as *bastard roofes* or false roofs; they were later called *ceiled roofs*, hence the word ceiling.

One can now imagine a void above the *bastard roofe*, along the whole length of the house, interrupted only by the principal truss and possibly the chimney. It was not long before these voids became used for additional accommodation and the traditional cottage became a two-storey dwelling. Not all of the house was made two storey; often the living room was left open to the underside of the rafters, presumably to prevent what must have seemed to many a rather confined and claustrophobic atmosphere below the low ceiling of the floor above. It must be remembered that the Great Hall was still very much a status symbol and to have at least one room of lofty construction would have been socially desirable.

THE EARLY ATTIC

Access initially to the attic space created was by a simple ladder and later by stairs often created from large solid timber steps supported on a crude sloping beam or carriage. Division of the attic into individual rooms was frequently achieved by filling in the spaces between the members of the principal truss to form a wall. The partition was usually created using wattle and daub (hazel sticks fixed between the timber members covered with plaster of lime mortar and cow hair). Some rural cottages dating back to the fifteenth century still retain this construction. Passage from one room to another through this partition is frequently somewhat restricted in height, the occupants having to step over the tie beam of the principal rafter and at the same time duck below the main beam. Figure 8.1 illustrates a simple attic partition.

The bedroom floor was usually supported by the bottom tie of the principal truss, thus reducing the floor joist span. Even with this reduced span, construction – presumably for economic reasons – often used undersized and relatively unseasoned timber, resulting in somewhat springy and uneven floors which would certainly not conform to today's design standards. The later development of fitting a relatively heavy plaster ceiling further exacerbated the floor deflection problem. Sawn elm or oak boards laid on top of the joist formed both floor and ceiling below, this construction resulting in the exposed beams and joists which are now considered an attractive feature in many early cottages. Figure 8.2 illustrates the floor construction.

FLOORS AND CEILINGS

The early floors were very basic, serving only to provide support for those above and their chattels. The quality of sawing being crude and the timber unseasoned, and with no jointing technique between the boards, there were large gaps between the boards; this reduced privacy between upper and lower floors.

These gaps not only reduced privacy; they also allowed heat from the room below to escape to the attic and created draughts. A simple ceiling was the next significant development, this being formed of timber laths, often not sawn but simply riven from freshly cut timbers into strips approximately 6 mm × 30 mm. The laths were nailed to the underside of the floor joists, leaving a gap between them of approximately 6–9 mm. Lime mortar, again mixed with cow hair, as a plaster was then applied to the underside of these laths, the plaster squeezing through the gaps to form a key above. Figure 8.3 illustrates a lath and plaster ceiling.

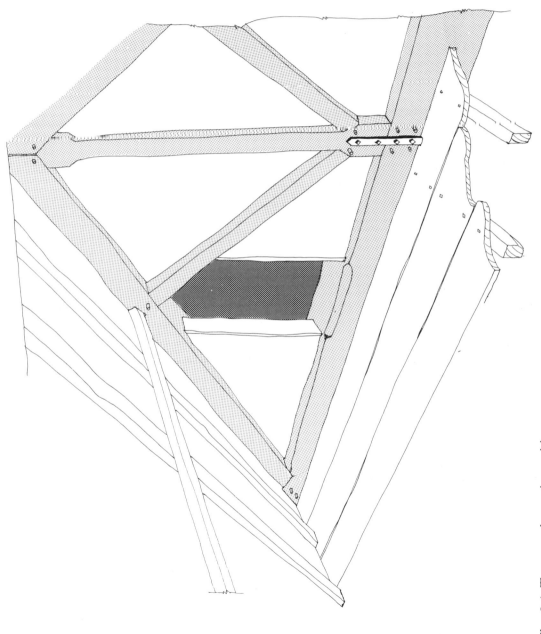

Fig. 8.1 Truss used as attic partition.

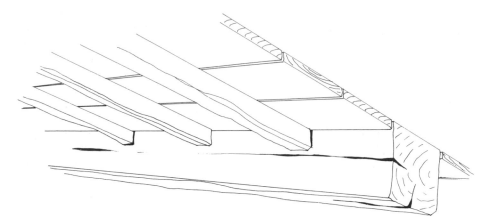

Fig. 8.2 Simple attic floor.

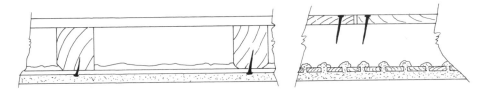

Fig. 8.3 Lath and plaster ceiling.

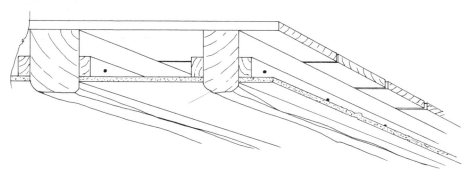

Fig. 8.4 Beamed ceiling.

Today, it is fashionable to strip off these old ceilings and fit a simple plasterboard ceiling between the joists, thus restoring the 'beamed' effect yet retaining both the privacy and insulation afforded by the ceiling. Figure 8.4 illustrates a typical construction.

COTTAGES

From these simple beginnings, dwellings developed both in size and sophistication of construction. The relatively cramped attic, although frequently used for lower-cost housing and servants' quarters of larger houses, became unpopular. The true second floor as we know it today, with its full-height ceilings, emerged as the traditional dwelling house. The full two-storey house became something of a status symbol, particularly in the more prosperous urban areas, but the one and a half storey cottage remained popular in rural areas where it seemed to fit more comfortably into the surroundings.

We should not lose sight of the fact that many rural cottages were supplied as tied homes by the wealthy landowner and they were obviously cheaper to build than a full two-storey structure. Development did, however, produce the form of dwelling shown in Fig. 8.5 which has the advantage of much improved headroom on the first floor yet does not have the full height or cost of a two-storey structure.

As roof tiles developed, roof pitches were lower thus reducing the attic space available and therefore its usefulness as a habitable room.

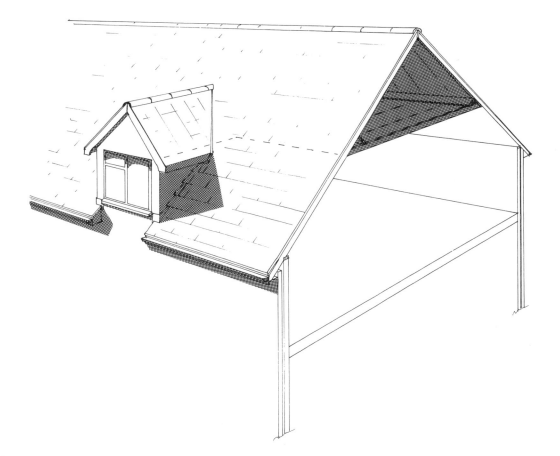

Fig. 8.5 Cottage attic.

WASTED SPACE

Improvements in general wealth and associated living standards resulted in most dwellings being constructed with two full-height storeys or even more, with the roof void being left for storage only. The use of interlocking pantiles and slates reduced roof pitches even further and thousands of houses were built during the industrial revolution with a roof space open from one end of the terrace to the other. Clearly in such situations access to the roof space was not desirable between dwellings for security reasons and the roof space was completely lost to the occupants. A typical terrace roof construction, albeit with a dividing or 'party' wall, is shown in Fig. 1.10.

After the Second World War, there was a great need to economise on the use of all imported building materials including timber, the main material used for the construction of dwelling roofs. TDA (Timber Development Association) prepared a range of standard roof truss designs which dramatically changed the domestic roof structure of the day, creating at one and the same time a structurally sound roof construction and one which used significantly less timber than pre-war constructions.

Chapter 4 deals with this type of construction, Figs 4.4 and 4.5 depicting the trussed roof structure.

THE FINAL BLOW

During the 1960s the trussed rafter began to be used for domestic roof construction; today it accounts for 95% of all dwelling roofs in this country. The trussed rafter placed at 600 mm centres reduces the usable roof void to a relatively small triangle in the centre of the roof (see Fig. 1.12). This area also has to accommodate the water tank platform, thus further restricting access to the roof void. The trussed rafter was certainly not designed to carry any floor loading, but adequately accommodates light domestic storage. Coupled with the introduction of the trussed rafter roof was the further development of double interlocking concrete tiles, which allowed even lower pitches down to 15° in some cases. Clearly, on relatively small spans the resulting roof void scarcely gave sufficient room to allow the installation and servicing of the cold water storage tank, let alone provide useful storage. Fortunately, current architectural style is towards a steeper pitched roof but only for aesthetic reasons, few housing developers taking advantage of the void so created to form additional habitable accommodation.

EXAMINING CONVERSION POSSIBILITIES

Before any firm plans or budget costs can be prepared, a survey of the existing roof void must be undertaken. The survey must be accurate; approximations even at this early stage may result in later inaccuracies in the drawings, some of which could mean costly changes to planned work at a later date.

Having surveyed the roof void, it is also necessary to survey the building below accurately, right down to ground level. It is most important to establish which of the internal walls are load bearing, i.e. which have a full foundation below them, in order that the existing supporting structure can be clearly established. In relatively modern houses it should be simple to acquire plans of the building, either through deeds, a builder (if known) or the local authority building control department which approved the building originally. On older buildings, a good guide is to lift floor coverings to find the direction of the floor boarding; the floor joists will run at right angles to the floor boarding and the walls onto which the floor joists are supported will be load bearing. Partition walls between rooms on the upper floors may not be of load bearing construction and this must be clearly established before considering them for possible support for the new floor to the attic.

TYPICAL ROOF TYPES

The terrace

The roof void in a typical terraced house is illustrated in Fig. 3.2; the distance between the party or compartment walls, i.e. those dividing the houses, is often relatively small and there are frequently load bearing internal walls to be found on the ground floor, supporting the first floor joists. As can be seen from the figure, by removing the 'collars' a useful room void would be achieved. The collar is, of course, there to tie the rafters together by stiffening the purlins and bracing the rafters. If the collar is removed some alternative restraint must be found. If room height is adequate, extra collars could be fitted to act as attic ceiling joists and to lighten the load on the purlins; an additional ridge purlin can be fitted, supported on steel shoes built into the party wall. It should be noted at this point that in some terraces no party wall will be found as has been stated earlier, and if this is the case then one must be constructed using methods adequate to provide both fire and sound performance to the adjoining properties.

Traditional hip and valley roofs

Traditionally constructed hip roofs will generally be of substantial construction, but they frequently defy engineering analysis to prove their stability. Traditionally, much reliance was placed on 'tosh' or 'skew' nailing, something the engineer is unable to prove but the passage of time has proven to work surprisingly well (see Fig. 8.6).

The hip void will be formed in a way similar to that illustrated in Fig. 3.7, and in most cases the purlins will be supported from a load bearing wall below. However, it is not uncommon to find the purlins actually giving little support to rafters except to bind them together; in such instances the rafters are actually the only support for the roof often using the hip rafters as main supporting members. To provide an adequate width to the loft room it may be necessary to consider removal of some of the purlins in these traditionally constructed roofs and this will be dealt with later in Chapter 11, 'Solutions'.

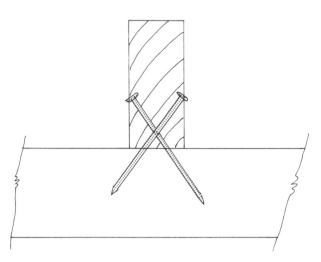

Fig. 8.6 'Tosh' or 'skew' nailing.

Lean-to or monopitch roofs

Although at first sight the lean-to part of some roofs may not be the most attractive to consider for conversion, it is often worthy of closer examination. To provide additional smaller room accommodation, the lean-to may well be one of the simplest roof voids to convert to attic accommodation (see Fig. 8.7).

Commonly at the rear of the property, the construction of the lean-to attic is unlikely to cause planning problems frequently associated with changes to the main facade of the dwelling caused by the construction of dormers in the roof. A larger dormer can often be created at the rear of the building thus affording much increased full height headroom as can be seen from Fig. 8.8. This gives the conversion of a relatively small roof void a much

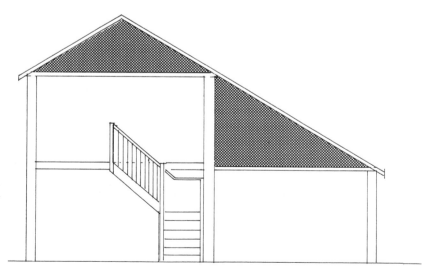

Fig. 8.7 Lean-to attic.

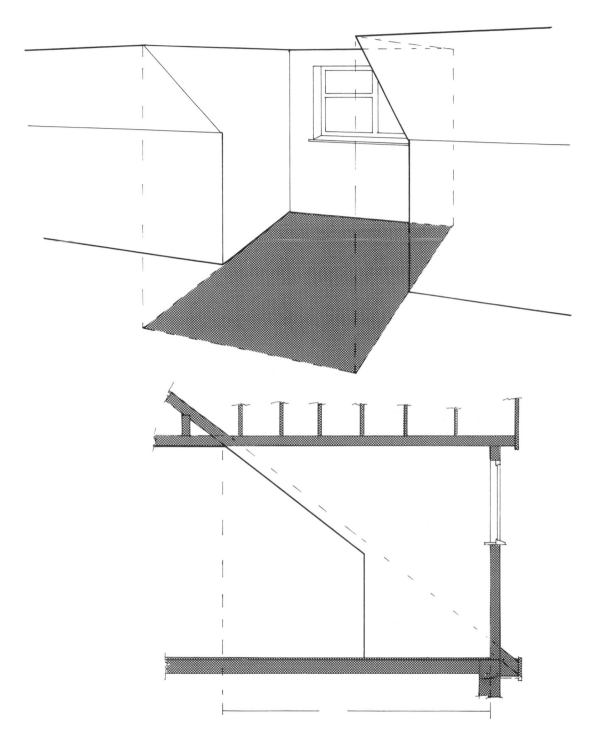

Fig. 8.8 Improved headroom.

better conversion value in terms of usable floor space. It is also unlikely that major structural problems will be encountered when attempting to convert a lean-to or mono-pitch roof unless, of course, it is constructed using trussed rafters. Simple lean-to roofs are often constructed using rafters supported off the wall plate at the higher level and lower level and with a single purlin between (see Fig. 8.9). Furthermore, unlike converting the main roof, the lean-to has one major advantage in that a separate access staircase is unlikely to have to be created. It is often quite simple to form a doorway into the new attic through the wall of the existing property from the first or second floor level (see Fig. 8.10).

The lean-to loft conversion will best work where tall rooms exist on the first floor, thus giving a good usable headroom in the attic. These tall rooms are often found in houses of the eighteenth and nineteenth centuries (see Fig. 8.11).

Bolted or TDA truss roofs

This type of roof has been described earlier and reference can be made to Fig. 4.4. Clearly the truss is a major component in such a roof construction and is not easily replaced if removed to provide a clear roof void for conversion. The only real solution is to provide purlins to support the rafters and ceiling joists, taking full account of the fact that the new attic floor will also have to be supported by some form of purlin as it is unlikely with this type of roof that there will be any supporting partition wall below.

To break the length of span of these purlins, it may be necessary to extend up a structural wall from the ground floor through the existing ceiling, the other supports for

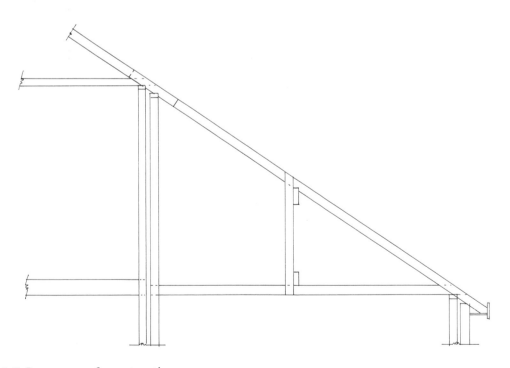

Fig. 8.9 Lean-to roof construction.

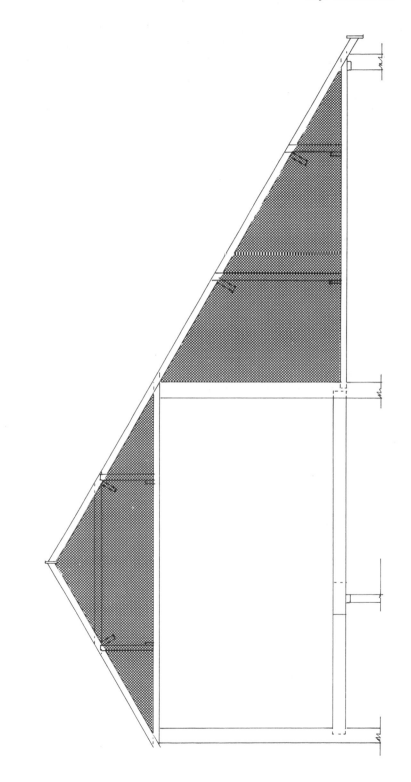

Fig. 8.10 Lean-to attic area.

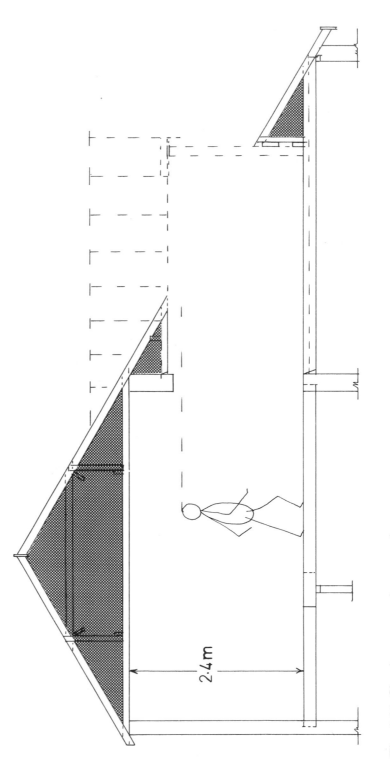

Fig. 8.11 Simple access to lean-to attic.

the purlins coming from the gable and brickwork. It must be *emphasised* that without engineers' structural designs, no element of the main roof truss should be removed before some replacement structure is in place.

Trussed rafters

This form of roof construction, although presenting a mass of small timber members within the loft, is no more difficult to convert than the TRADA type noted above. The principle of the roof is similar in both cases, i.e. to provide a clear spanning truss avoiding the need for load bearing internal partitions. Unlike the TRADA truss the timbers in the trussed rafter will be of much smaller cross section and there may therefore be a need for more pick-up or purlin points if internal members are to be removed to create the attic. Again, it must be *emphasised* that no member of the trussed rafter roof should be removed without an engineer's design replacement structure in place (see Fig. 8.12).

LETTING IN LIGHT

This initial examination of the loft structure is not complete without consideration of whether light might be provided to the roof construction by means of dormer windows, roof windows or roof lights. Obviously on a terraced house there is not the possibility of cutting windows into the gable walls, except of course if you are fortunate enough to be at the end of the terrace. Consequently, dormers or roof lights will have to be used to provide light and ventilation to the loft conversion. These may have to be located in such a way as to take into account the structure of the roof, yet show awareness of their impact on the external architecture of the house. This subject will be dealt with in more detail later (see Fig. 8.13). Refer to Chapter 3 for other dormer shapes.

One of the most efficient means of lighting attics is the roof window. There are a

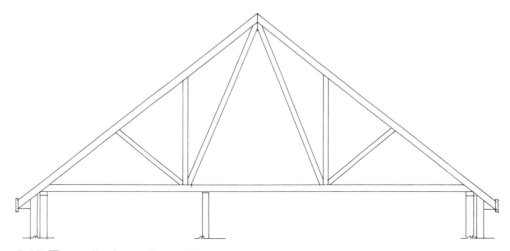

Fig. 8.12 Trussed rafter attic problems.

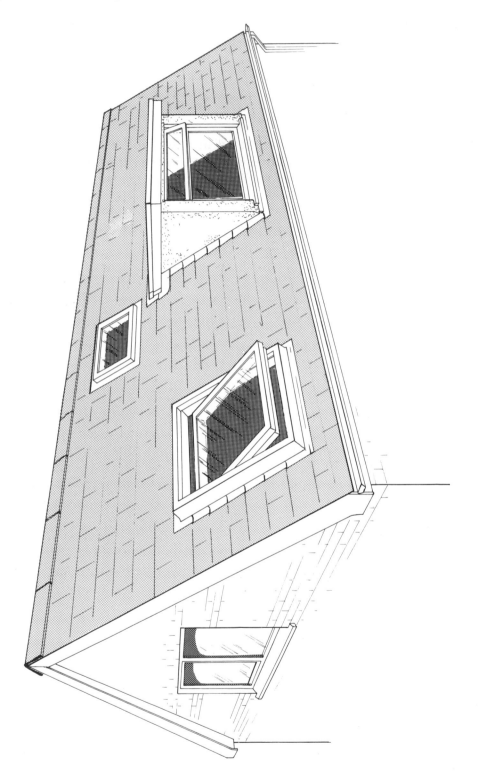

Fig. 8.13 Letting light into the attic.

number of proprietary manufacturers of roof window available to suit most pitches from virtually a flat roof to the side of a mansard roof which would frequently be 60 degrees pitch or more. These windows can be provided with sophisticated opening systems giving trickle ventilation, double glazing and incorporate various types of sun blind equipment. Designs are available from simple small windows for bathrooms and toilets to combination sloping and vertical windows which offer the ability to allow considerable amounts of light and ventilation into the loft.

Some typical installations are indicated in illustration 8.14. The left hand sketch indicates roof windows fitted between trussed rafters, thus avoiding any structural disturbance to the roof itself and offering probably the most convenient and cost effective way of introducing an opening roof window into the roof space. The centre sketch illustrates a typical installation involving trimming an opening which has been described elsewhere in this book, thus allowing a larger roof window to be installed. The right hand illustration shows a horizontally split window running from near ceiling height to the floor level of the loft, and this type of roof window is available either with an opening upper and fixed lower section or as a roof 'balcony', in which the lower section pushes out to a vertical position and with the upper section opening upwards from its top hinges allowing the occupant to effectively walk out into the open air, with protective guard rails automatically locating themselves for safety reasons.

CAN WE STAND UP?

Until the introduction of the Building Regulations 1985, there had been a minimum height requirement for rooms in roof spaces, but these no longer exist. Depending on the intended use of the attic, it is desirable to have the maximum area of full height attic possible even if it is only just over 2 metres high. Clearly the pitch and span of the roof will affect the usable room within the attic; a narrow but steep pitch roof may be as useful as an attic conversion as a larger span, shallow pitch roof. In this respect many houses of the 1950s, 1960s and 1970s will not prove very easily converted because of the trend in that period to lower pitch roofs. There is also regional variation in pitch preference, but in recent years the trend has generally been for steeper pitches. As can be seen in Fig. 8.15, the low pitch roof is one of the reasons we often see large, somewhat ugly, dormers with their troublesome flat roofs extending almost to the ridge just to afford adequate internal headroom within the attic conversion.

VALLEY

The valley, if to be included in the proposed attic, is not a construction area to be tampered with if at all possible. Access within the attic will therefore have to take full account of this feature, and this will have an effect on the planning of landing and corridor areas between attic rooms. It is, therefore, advisable if possible to design the valley into the landing area keeping it out of the habitable rooms where better headroom is desirable (see Fig. 8.16).

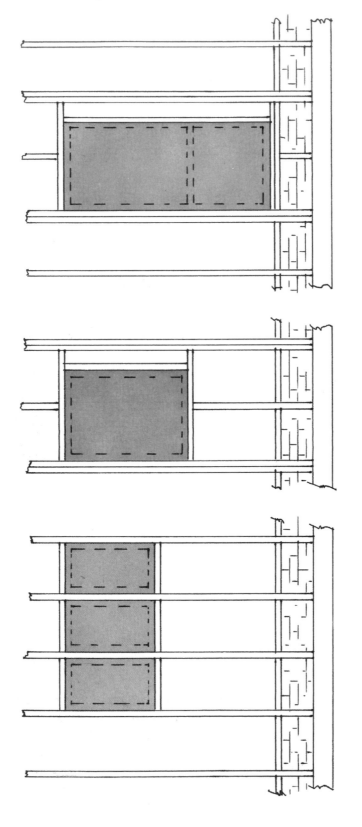

Fig. 8.14 Some typical roof installations.

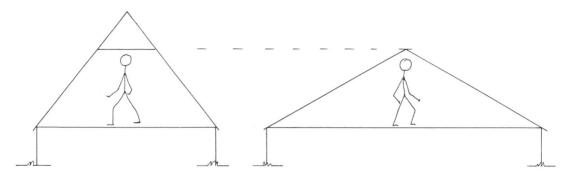

Fig. 8.15 Headroom problems.

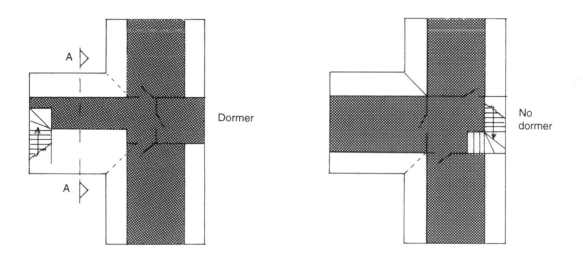

Dormer

No dormer

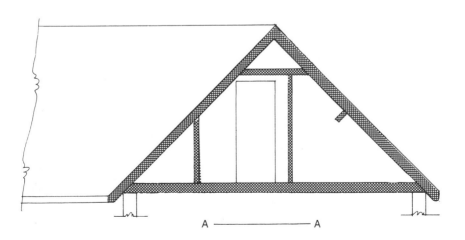

Fig. 8.16 Effective attic layout.

ACCESS TO THE ATTIC

It is quite easy in the early stages of planning the additional space to be gained from the attic conversion, to overlook access to the attic from the rooms below or in some cases adjacent, as in the case of a lean-to attic conversion. The staircase or landings and possibly passages will take up an unexpected and in some cases a disappointing amount of the valuable attic floor area conversion. Whilst the Building Regulations have been relaxed somewhat in their requirements concerning the staircases to attics, allowing the loft 'ladder' type staircase and of course the more decorative, smaller spiral staircases, it must not be forgotten that good access is needed not just for personal reasons but also to install furniture and fittings.

The small spiral staircase may provide a very interesting architectural feature within the house and indeed be very compact within its plan space requirements, but is not very practical for furniture or even for such items as baths if these are to be installed in the new conversion. It is *not* a very clever idea to install the bath and then fit the staircase! This is the same as fitting a large water storage tank in the loft and then making the trap door too small to get a replacement through.

It is strongly recommended, therefore, that a sensible sized staircase be installed for both access and safety reasons, and its impact upon the ground below must be considered. It may, for instance, be necessary to sacrifice a small ground floor room to provide space for the staircase and this in turn may have an effect on the planning requirements of the loft space above. The stairwell, whichever form it takes, will necessitate a reasonably large hole to be formed in the floor of the new loft and the structural implications of creating this must be carefully planned.

THE LAST RESORT

If all looks lost as far as a conversion of the existing loft structure is concerned for reasons of:

- excessive span for new purlins,
- inadequate headroom and usable room width,
- the need for too many large dormers,
- problems with attic floor support,
- a defective existing roof structure needing expensive repairs, or
- the need to re-tile the existing roof,

then consider a complete roof replacement as it may not be so dramatic or excessively costly as may at first be thought. It may also be considerably cheaper than the alternative possible extension to the property on the ground floor even if plot space and planning constraints would allow. We will deal in more detail with this possibility in Chapter 11, Solutions.

CHAPTER 9

Obligations – Visual Impact on Your Home

Any loft conversion creating living or sleeping accommodation within the roof space will of necessity need some form of window for light and ventilation. Although you may escape these requirements for a simple bathroom or toilet, even in these latter cases the need for ventilation by extractor will result in some form of additional vent pipe or grille in the roof or gable end.

Before dealing with specific requirements concerning planning approval for your proposed conversion, the likely impact of the alterations on the architecture of your house must be carefully considered. The type of window installation – dormer, roof window or roof light – will depend very much on what is seen to fit in architecturally with the style of your building. It would hardly be appropriate, for instance, to install a steep-pitch dormer with a small gable on a relatively low pitched building.

Conversely, flat roof dormers will look totally out of place on a steep-pitch cottage roof (see Fig. 3.17 for dormer styles and Fig. 8.13 for further examples).

To gain the maximum amount of full-height room area within the attic, there is a temptation to settle for very large dormer windows which by their very size will have to carry flat or monopitch roofs; if this is the only solution then they should be kept to the rear of the property where they may well prove more acceptable to the planning authorities, simply because they will bring about no major visual change to the character of the main facade of the building. The degree of change to the building acceptable to the planning authorities will be dependent very much upon the building location. For instance, a loft conversion proposed in a house on a small development of architecturally similar dwellings will be viewed differently to the same proposal on a similar individual house in a street of mixed architectural styles or on an isolated country cottage.

Every effort, therefore, should be made to design new roof architecture which is in sympathy with the style of the existing property and that of the surroundings. The fenestration, i.e. location of windows in the facade, of the existing building will almost dictate the lateral location of dormers and roof windows, and this in turn may have some effect on the positioning relative to the rooms created within the attic.

Figure 9.1 illustrates unsympathetic dormer location, possibly decided by a desire to locate the window centrally in the rooms converted within the attic. This dormer, in addition, has a flat roof which is not in keeping with the style of the house; and the style of the window itself, whilst it may be regarded as desirable by the occupant, is not in keeping with the style of the existing windows. If the age of the house means that standard windows of that style are no longer available then special windows should be purchased to match the existing.

Figure 9.2 presents a better facade. There are numerous examples of loft conversions in which relatively cheap, double glazed, large paned, UPVC windows have been fitted in Georgian- or Victorian-style houses, and do nothing for the character of the dwelling, probably reducing the value of the property.

It may have been detected that people are receiving considerable encouragement here to consider professional architectural advice at this relatively early stage. Indeed it is strongly recommended. It is unusual for the layman to be able to appreciate and produce *style*, and it is worth employing a professional, but do find an architect who can show examples of loft conversion work already undertaken. Use the architect for planning only at this early stage of the proposed conversion as it may well be better, if planning approval is obtained, to employ a qualified builder to examine carefully the structural implications of the proposed loft conversion. Again, select a builder with great care – one who has a proven track record on loft conversion work. With the greatest respect to architects, whilst many can produce an aesthetically pleasing proposal, they are not necessarily the best profession to make practical and therefore economical proposals where major structural alterations to a roof construction are involved. Ideally, a combination of both professions should give the best results.

THE PLANNING APPLICATION

Whether or not the work is entrusted to a professional architect, an understanding of what is required in the preparation of the planning application is useful.

The Town & Country Planning General Development Order 1988 came into effect on 1 May 1989. This order significantly relaxed the mandatory obligations on certain developments on residential buildings. Schedule 2 of the order widened the circumstances in which extensions can be exempt from planning approval in England and Wales. Different rules apply in Scotland.

Whilst Class A of Part 1 concerns itself with general extensions, Class B is concerned with dwelling enlargements consisting of addition or alteration to a dwelling's roof.

The following is not permitted under Class B:

(1) Any alteration resulting in the increased height of the existing roof.
(2) Any new work which extends beyond the plane of any roof slope which fronts a highway.
(3) Any alteration which increases the cubic content of a terraced house by more than 40 m^3 or any other house by more than 50 m^3. If, however, other parts of the building are being extended as well as the roof, then rules 4 and 5 also apply, taking into account the whole volume of the dwelling.

Fig. 9.1 Poor dormer design.

Fig. 9.2 Good dormer design.

(4) Any alteration which increases, in the case of a terraced house, the cubic content of the original house by more than 50 m^3 or 10%, whichever is the greater, and, in the case of any other type of house, by more than 70 m^3 or 15%, whichever is the greater.

(5) Any alteration which increases the cubic content of any dwelling by more than 115 m^3.

(6) Any alteration to a dwelling which is on:
  (a) National Park land,
  (b) land designated as a conservation area under Section 277 of the Act,
  (c) land controlled by the Secretary of State and Ministry of Agriculture and Fisheries for the purpose of enhancement and protection of the natural beauty and amenity of the countryside, under the Wildlife & Countryside Act 1981.

Alterations to a roof resulting in a material alteration to the shape of the building are not covered by permitted development.

Different local authorities interpret the rules in different ways, and it is therefore well worth seeking the advice of the local authority planning officers concerning particular proposals and their effect upon the existing property. To proceed with even the preparation of drawings and specifications on the assumption that the proposed alterations to the roof structure may be exempt could prove costly. Local authorities are empowered to enforce reinstatement to the original if subsequent approval is refused. It is likely that the planning officer's advice will be free; the formal planning application cost is not excessive and would be money well spent if it forestalled a situation in which work was carried out and then identified by a local authority and subsequently refused.

If the proposed loft conversion work falls outside that permitted in Part B, then full planning permission must be sought.

THE STRUCTURAL IMPACT ON THE HOME

At this juncture the aspect of control on loft conversion is the Building Regulations. It is often thought by the layman that exemption under planning will mean freedom to convert; another fallacy is that once planning approval has been granted, there are no other obligations under the law and work may proceed. Planning controls concern themselves only with the visual impact on the local environment of the proposed conversion, whilst the Building Regulations are concerned with safety, health and welfare. Building Regulations approval is required for all structural conversion work to any building. Building Regulations approval must therefore be sought as the Regulations take into account the following aspects of the conversion:

- Structural stability;
- Heating and thermal performance of the converted structure;
- The installation of heating appliances;
- The fire performance of the structure, particularly if the conversion creates a third storey to the building;
- Ventilation of the habitable rooms created;
- In some cases, means of escape from the created attic rooms;
- Considerations for soil and waste drainage if a bathroom, w/c or shower is installed;
- Consideration of roof rainwater run-off if the roof structure is materially changed.

CHAPTER 10

The Conversion

MAKING A START

Having prepared rough plans to examine the roof structure's potential, the next step is to prepare detailed plans and a specification and this must include a survey of the structure below to assess its load bearing capacity.

THE SURVEY

Obviously, accurate measurements of the existing building are important but attention must be paid to the construction of walls, floor and roof. If drawings and specifications do not exist for the property to be converted then the task of recreating these must be faced, including establishing foundation location, size and depth.

It is worthwhile examining what exactly is desired of the conversion; for instance, if it is intended to create storage only rather than habitable accommodation then the loading on the floor below and consequently on the structure below that, may well be significantly different.

The aspects to be carefully considered are as follows:

(1) The additional load of the new floors;
(2) The additional load of new walls;
(3) The additional load of new ceilings;
(4) The additional live load of the occupants and the fittings;
(5) *In the case of a third floor*, the additional weight of fire protection to walls around stairwell;
(6) *In the case of the creation of flats*, the additional weight of the fire-resisting floor;
(7) *If the conversion is to be used as a home-based office*, the additional weight of office equipment and storage of files.

Any loft conversion which includes habitable accommodation is likely to add around 30% to the existing foundation loading, much of this being transferred to the foundations via lintels over windows and doors which will probably only have been calculated to carry a conventional load from the roof. Consequently, any conversion will have a significant effect on the structure below and as most bungalows have the same external load bearing wall construction as a two storey house, there is unlikely to be a problem with the wall itself; attention will have to be paid, however, to such items as the lintels, as explained above.

Checklist for survey

- Location of foundations of existing external walls and load bearing partitions;
- The type of wall construction for existing external walls and partitions;
- The type and specification of existing lintels over windows and doors;
- Direction and size of floor joists on the ground floor, if any, thus helping to establish which partitions are load bearing;
- Direction, size and spacing of existing ceiling joists;
- Size and spacing of existing rafters and other roof timber components;
- Accurate dimensions of the existing building – not just on plan but all elevations likely to be affected by the conversion;
- Accurate heights of all rooms, thicknesses of ceiling zone, shape of roof space and locations of any purlin supports that may be in the existing roof structure;
- Accurate location within the plan of any services in the loft area, such as chimney stacks, soil and vent pipes, water storage tanks and all service pipes. It must be remembered that some or all of these may have to be moved as part of the conversion.

PLANS AND SPECIFICATION

From the above survey produce a full set of drawings to include:

- Scale plan not less than 1:50 scale;
- Scale elevation not less than 1:50 scale;
- Scale section or sections; as more than one section may be necessary if the roof shape and/or its construction changes from one part of the building to another, this should be prepared to as large a scale as possible.

All of the above to be prepared on a paper or negative which may be conveniently copied for reproducing the drawings. It must be remembered that planning and building regulations consume large numbers of copies of the plans, and furthermore additional copies will be required to obtain quotations from building and/or sub-contractors for the various trades involved. Do not be surprised if, by the time the conversion is completed, 20 copies of all drawings have been produced!

Obviously the next step, using the existing floor plan of the floor below the proposed conversion, is to sketch on tracing paper laid over the plan the proposed loft conversion plan. This will immediately show what is not practical and at this very early stage one of the first things to be considered is access to the loft space.

Having produced a detailed and accurate set of drawings of the building, as existing, and of the building as proposed in the new conversion, a fully detailed, descriptive specification must be prepared even if it is proposed to carry out the work on a DIY basis. The specification will be even more essential if you intend to employ a builder to carry out the project either for yourself or for a client, and the specification is particularly useful when co-ordinating subcontractors' works. A poorly detailed, hurriedly prepared specification will lead to inaccurate and incomplete quotations or tenders for the work involved and will inevitably result in many extra costs as work proceeds. This in turn, generates ill will between the parties involved, and ill will generally results in poor quality work. As far as the building owner is concerned the worst result of an ill-prepared specification and drawings is the additional, unforeseen costs which will not have been budgeted for and could, if significant, cause financial difficulties by the end of the contract.

WRITING THE SPECIFICATION

It is advisable to start the specification with a detailed description of the work involved; this is known as 'scope of works'.

When starting to write the specification of the works, it is advisable to describe the works at the very lowest levels of intended alteration which may, in some cases, mean modifications to foundations and will contain descriptions for:

- Providing temporary support for floors and existing ceilings in order to strengthen or replace lintels;
- Preparing, probably, even at this early stage, the stairwell which will afford access to the works in the intended conversion area.

Specify precisely; leave nothing to question by the builders or even yourself at a later date at the time of construction, thinking through and describing all of the building operations from the start of the element being considered to its natural conclusion of the reinstatement of surfaces and decorations. An example is given below for the replacement of an existing lintel with a new lintel capable of carrying a new floor load above:

'Window W4: carefully remove carpets and provide dust-proof screen on dining room side of window. Provide temporary structural support to ceiling and roof above by fitting a temporary purlin in the roof space above over the areas of the rafter currently supported by the lintel. Purlin to be supported off struts directly on bearers onto ceiling joists.

Fit a spreader beam immediately beneath the struts under the existing ceiling and support on Acrow or similar props. The Acrows to be set on a bearer on the floor. When all load has been taken from the lintel by the propping system, carefully remove the existing lintel, replace with new lintel and bed solidly. Do not remove props until mortar has set. Not less than 48 h after the replacement of the lintel, prop system can be removed and the area restored, made good and decorated as necessary.'

Avoid vague and incomplete specification clauses such as, 'Fit new floor to attic space.' This should be specified as follows:

'Provide and fix 19 mm tongued and grooved jointed, V313 flooring grade chip-board conforming to BS 5669:1989 Type C4. Board to be fixed with 63 mm long annular ring shanked nails, including gluing all tongued and grooved joints. 38 × 50 mm softwood noggings to be used around the perimeter of the room to support flooring and at any square edge joint in the board.'

This will avoid:

(1) the supply of inferior boarding;
(2) falling below the requirements for a V313 moisture-resisting grade chipboard in bathroom areas;
(3) the irritating problem of squeaking floors caused by incorrect nails and incorrect jointing of the boards;
(4) the flexing of the floorboard between floor joists through inadequate support.

The result is, of course, a good sound job at the price quoted, and if any of the specified items have not been carried out at the time of inspection, or a badly creaking floor becomes evident after use, then the employer has a legitimate claim against the builder. The moral of all this is to spend time in the preparation and planning and you will save hours if not days of delay and frustration later. Remember, if you are the architect or builder, that your client may well have to remain in occupation throughout the conversion and the minimum of disruption is therefore desirable. Delays caused by unspecified and therefore unforeseen work will not easily be tolerated at this time of great disturbance to the household.

IMPACT ON OCCUPANTS DURING CONVERSION

In the same way that a surgeon advises his patient on the likely effects on the body after the operation, in terms of likely recovery time, convalescence time and discomfort to be expected immediately following the operation, not forgetting, of course, the benefits in the long term; so too must a designer or builder advise his client (if DIY then fully brief the wife!) exactly what will be involved in the way of disruption to normal family life, discomfort caused by the noise of building and inevitably traffic through the house, storage outside the building and dust. The work should be explained in great detail as many lay-people cannot fully comprehend drawings and may not understand exactly what is planned. It is advisable to walk them around the ground floor, point out such things as that, perhaps, the study has to disappear to become a stairwell and that a window may have to be bricked up. Explain that the bathroom is to be moved from the ground floor to the attic, creating a study. The client may not have appreciated that this would involve the removal of a WC from the ground floor, resulting in excursions to the attic for this convenience.

Although it may be difficult to get the occupant(s) into the loft space before conversion, it may be worthwhile in an attempt to explain more fully the layout of the proposed rooms in the conversion and the access point to them. It is particularly important to explain that the side wall height in the attic may only be 1.2–1.5 m, some people being surprised to find sloping ceilings when the conversion work is completed. They shouldn't, of course, if this is fully explained beforehand. Try to explain the likely time for the works and at this point consider the time of year in which the operation is to be carried out. If there is no urgency, then it is clearly an operation better executed in spring and summer when it is easier to tolerate short-term loss of heating which may occur during conversion of the heating system.

If a bedroom is to be converted on the ground floor for access to the attic, then alternative sleeping arrangements may become necessary if, for some reason, the new attic bedroom cannot be created in advance.

BUILDERS' STORES AND MATERIALS

Many materials will be needed for the building work, so where are they going to be stored? Consideration will have to be given to allowing the builder (if space is available) to bring in a store/small office, or possibly to vacating the garage to provide this accommodation. Having done this, where is the car to be parked whilst builders' trucks are in and out? If these things are not thought through it will be a considerable shock when a builders' merchant's lorry turns up and starts to crane packages of wood and sheet materials onto the garden.

It will undoubtedly be necessary to have one of the rubbish skips now so commonly seen parked in or near the premises, and this may need to go just where you were considering parking the car, having removed it from the garage! Lorry access for deliveries and for the builder will need to be maintained at all times during working hours. Safety lighting may be necessary at night if an obstruction is left in the highway overnight.

Finally, because of the hammering and the activity in the loft space during the creation of the attic, it is almost inevitable that some cracking of ceilings, and ceiling-to-wall joints will occur in the building below, making redecoration necessary. Dust also, however careful the builder is in providing dustproof screens, will penetrate into the whole of the building. Thought must therefore be given to the probability of having to redecorate at least the ceilings of the building below.

CHAPTER 11
Solutions

This chapter examines ways of converting three of the main roof construction types and also roof replacement as a possible solution to the problem of providing additional accommodation in the roof space. In all cases the author has assumed that the roof structure is sound and that repairs are not necessary. Should this not be the case then these will be the subject of separate careful inspection and specification for the work. It is assumed that roof coverings are also sound and that structural supports have been checked and are capable of carrying the additional dead loads (those from the structure itself) and the live loads of furniture and occupants.

The roof types to be examined are:

(1) The common roof (see Fig. 1.10);
(2) The bolted truss roof (see Fig. 4.5);
(3) The trussed rafter roof (see Fig. 5.5).

Having examined the above roof types, consideration will be given to the possibility of a replacement roof structure which may be necessary if the roof shape is totally inadequate to provide usable space, or if the roof structure is in such a bad condition that repair is considered to be uneconomical. In all cases, a simple gable-to-gable straight ridge roof will be considered.

BASIC ACCOMMODATION

Figure 11.1 indicates basic requirements to be achieved:

- Establish room height required, not forgetting that new floor joists will effectively raise the floor level above that of the existing ceiling joist level in the loft.
- Establish over what width of building the room will have practical use, bearing in mind the side wall height regarded as tolerable.

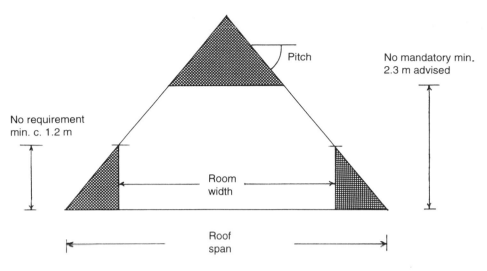

Fig. 11.1 Basic requirements of an attic.

THE COMMON ROOF

In Fig. 11.2, a very useful room can be provided but all internal structural components will have to be removed or replaced with components in different locations. Purlins A may be acceptable and could possibly be left exposed in the loft space but if they are at right angles to the rafter, they are probably doing little to support the structure and they may (depending on the age of the building) be ineffective owing to inadequacy in size. It must be remembered that prior to 1930/40, little structural design was applied to roof

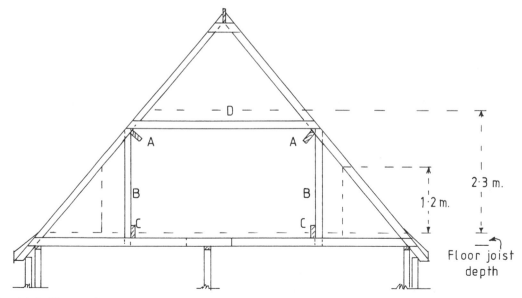

Fig. 11.2 Converting the structure.

structures except that of 'rule of thumb', and whilst some roofs may have oversized components many would not satisfy the structural engineer today. Reference should be made to the Building Regulations tables to ensure satisfactory purlin sizing or full structural design.

Hanger B, ceiling binder C and collar D are all clearly obstructing the attic and will have to be removed. The proposed solution is indicated in Fig. 11.3. *Before removing any existing components the new structure must be installed and existing rafters and joists securely fixed to the new components.* Purlins A1 and A2 are to be installed complete with a new collar which acts as the attic ceiling joist at B. If possible, when installing the purlin, the actual bearing should be dry, i.e. not supported on a mortar bed as mortar takes a considerable time to reach full strength, thus making some temporary support necessary. Bedding dry on a bitumen-type DPC will mean that the purlin can be loaded immediately and any brickwork making good around the purlin is non-structural. New Hanger C acts as attic room side wall studs and continues to carry the ceiling joists of the original roof structure stiffened by binder D. The new floor joists E will be installed between the existing ceiling joists (see Fig. 5.15).

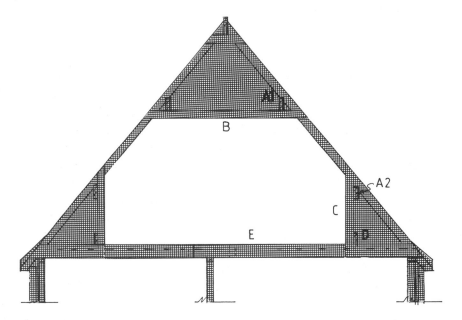

Fig. 11.3 Converted structure.

BOLTED TRUSS AND TRUSSED RAFTER ROOFS

Unlike the common and purlin roof, both of these roof types are 'tied' together by the ceiling joist of the truss members; this tying keeps the roof together. Without the tie, significant side-thrust would be imposed upon the supporting external walls of the house which carry the rafters – a side thrust for which they are not designed and may not

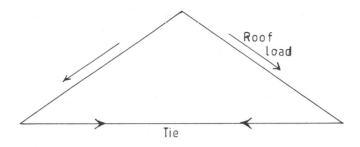

Fig. 11.4 Stresses in roof structure.

be capable of supporting. Figure 11.4 indicates the direction of stresses in the roof structure.

The internal members simply offer intermediate support for the rafter and ceiling joist of the truss, thus allowing a reduction in timber size. It is, then, essential to leave these internal members in place until the new structure has been established. As these roofs are engineered structures, no attempt will be made here to give 'rule of thumb' advice on internal truss member replacement – this advice must come from a structural engineer.

As both the bolted truss and the trussed rafter roofs are generally designed to span from external wall to external wall without intermediate support, the provision of a new joist capable of carrying the floor is going to be difficult simply because of the span involved. One possible solution is illustrated in Fig. 11.5. The glued laminated beam purlin illustrated will have to span between load bearing walls running across the building or from gable end to gable end. The purlin will be designed to carry both floor and roof structure loads via the side wall studs indicated in the illustration.

A similar structure is illustrated in Fig. 5.17. Although this particular illustration assumes a prefabricated rafter structure, clearly the beam and floor joist system could be applied to the conversion. Similarly, the upper roof structure of the trussed roof will also need supporting on some form of heavy purlin, again as indicated in Fig. 5.17, but in this case the rafter would be the existing rafter. Some of the internal members may be those of the existing trussed rafter or bolted truss roof with new ceiling joists being fitted on metal hangers supported on the new purlins.

It will be noted from the above brief description that on both the TDA and the trussed rafter roofs no attempt has been made to give a detailed description of conversion because this must be designed by a structural engineer. In both cases the basic rafter and ceiling ties have been left undisturbed.

ROOF REPLACEMENT

With attic trussed rafters, prefabricated in a factory (now so commonly and economically available), a roof replacement is a much simpler and quicker operation than may at first be thought. Most trussed rafter manufacturers can provide a 'whole' roof design

Fig. 11.5 Purlin 'wall'.

service, supplying not only the individual trussed rafters but also any girder trussed rafters, probably pre-nailed in the factory, together with all the necessary infill timber work and connecting light steel fabrications. Properly organised and planned, the stripping of the existing roof and the installation of an attic roof structure by crane is probably no more than a 2-day operation. If possible, the existing ceiling of the building below should be left in place, perhaps with some temporary support, and a contingency for weatherproofing will undoubtedly have to be considered.

If it is not the intention to use a crane, it may be practical to dismantle part of the existing roof and replace it with the attic structure, providing felt and battens to the new attic area and thus reducing the temporary weatherproofing problems which will inevitably occur. Clearly, the roof replacement option is one which is best carried out at such times as before somebody moves into a new home, or in a building which can at least be temporarily vacated. Using this particular option, it will probably be practical to construct the new attic structure, completing the installation of dormers, weatherproofings, retiling, making good or extending up gable ends and carrying out much of the internal work before penetrating through the existing ceiling into the building below. This, of course, will keep disturbance to a minimum.

VENTILATING THE ROOF VOID

In creating the new attic, three new roof voids will probably have been formed and these should be ventilated, particularly now that the majority of the roof is to be heated, thus causing possible condensation problems in the now much smaller voids. Various forms of ventilation at the eaves are detailed on page 150 *et seq* and the subject is covered in

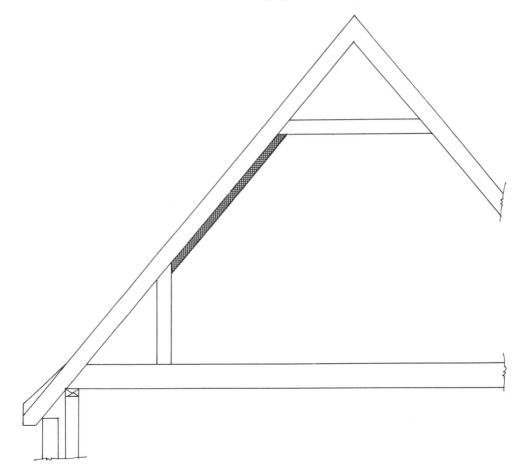

Fig. 11.6 Insulation and ventilation.

more detail in Chapter 7, the attic being dealt with on page 152. The area for particular attention is that of the sloping ceiling which, if the existing rafters have been left in place, may not provide sufficient depth to install adequate insulation and at the same time maintain a ventilation void above. It may be necessary to provide some additional depth to the rafter by fixing a packer beneath, creating the construction indicated in Figs 11.6 and 11.7. It is necessary to maintain the ventilation space above and to do this either plywood can be installed or, if the roof covering has been stripped, it may be possible to introduce some of the proprietary plastic components for this purpose.

On the question of vapour barriers, it is worth repeating that in the author's opinion vapour barriers in the attic are more trouble than they are worth, provided the roof voids have been correctly ventilated. A vapour barrier does exactly what it says: it prevents vapour penetrating through into the structure beyond. This of course means that it stays within the room and if the rooms are not continuously heated or are subject to high humidity levels (e.g. bathrooms and showers), this vapour cannot escape and may well condense within the lining material. This in turn, particularly in the areas of dead air circulation, can rapidly give rise to unsightly mould growth which is also difficult to eradicate once established. By dead areas of the room is meant those corners which are not subject to particularly good ventilation.

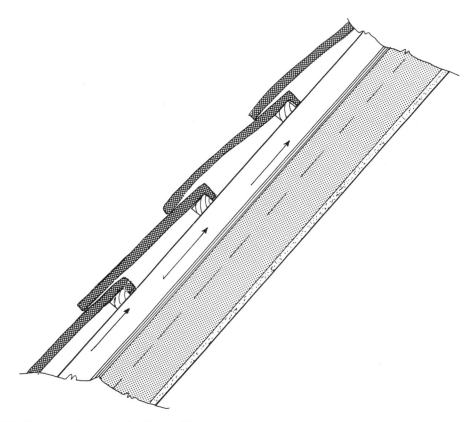

Fig. 11.7 Section through sloping ceiling.

OPENINGS FOR DORMERS AND ROOF WINDOWS

Guidance on this is given in the final pages of Chapter 3 and in Chapters 4–7.

CONCLUSIONS

Chapters 9, 10 and 11 have attempted to give some guidance on roof structure conversion to living accommodation. The information is not exhaustive and does not cover all forms of roof structure or envisage all forms of intended use for the attic. It is not difficult to imagine a conversion for a snooker room with the installation of full-size slate bed snooker table, for instance, which would clearly introduce a whole new set of parameters for the loading of the new floor. Similarly, the installation of a grand piano, an exceptionally large bath or even the infamous water bed can be envisaged.

The imposed loads from a newly created private library in the attic would also be an unusual structural problem to solve. These chapters are intended to attract the attention of the DIY enthusiast or designer embarking on a proposed conversion, encouraging him to consider not just the conversion but its impact upon the occupants, the structure below and the architecture of the completed building. In the author's opinion, the key to a successful conversion can be summed up as follows:

(1) Careful and detailed planning of the proposed conversion;
(2) Careful counselling of the DIY enthusiast's family or the architect's client on the impact upon them during the conversion and the benefits following the completion;
(3) Full appreciation of the impact upon the architecture of the building;
(4) A fully detailed costing of the works before embarking on the project.

BIBLIOGRAPHY AND REFERENCES

REFERENCE AND CONSULTATIVE DOCUMENTS

BS 1282: 1975	Guide to the choice, use and application of wood preservatives.
BS 1579: 1960	Specification for connectors for timber.
BS 4072	Wood preservation by means of copper/chrome/arsenic compositions.
BS 4072: Part 1: 1987	Specification for preservatives.
BS 4072: Part 2: 1987	Method for timber treatment.
BS 4169: 1988	Specification for manufacture of glued–laminated timber structural members.
BS 4978: 1996	Amd 9434, April 1997. Specification for visual strength grading of softwood.
BS 5250: 1989 (1995)	Code of practice for control of condensation in buildings.
BS 5268	
BS 5268: Part 2: 1996	Amd 9451. Structural use of timber.
BS 5268: Part 3: 1998	Structural use of timber. Code of practice for trussed rafter roofs.
BS 5268: Part 4	Fire resistance of timber members.
BS 5268: Section 4.1: 1978	Amd 2947 and 6192, March 1990. Recommendations for calculating fire resistance of timber structures.
BS 5268: Section 4.2: 1990	Recommendations for calculating fire resistance of timber stud walls and joisted floor constructions.
BS 5268: Part 5	Code of practice for the preservative treatment of structural timber.
BS 5268: Part 6	Code of practice for timber frame walls.
BS 5268: Section 6.1: 1996	Amd 9256, June 1996. Dwellings not exceeding four storeys.
BS 5268: Part 7	Recommendations for the calculation basis for span tables.
BS 5268: Section 7.1: 1989	Domestic floor joists.

204

BS 5268: Section 7.2: 1989	Joists for flat roofs.
BS 5268: Section 7.3: 1989	Ceiling binders.
BS 5268: Section 7.4: 1989	Ceiling joists.
BS 5268: Section 7.5: 1990	Domestic rafters.
BS 5268: Section 7.6: 1990	Amd 6902, Feb 1992. Purlins supporting rafters.
BS 5268: Section 7.7: 1990	Purlins supporting sheet or decking.
BS 5534: Part 1: 1997	Code of practice for slating and tiling (including shingles).

BS 6399

BS 6399: Part 1: 1996	Code of practice for dead and imposed loads.
BS 6399: Part 2: 1997	Code of practice for wind loads.
BS 6399: Part 3: 1997	Amd 6033, 9187 and 9452, May 1997. Code of practice for imposed roof loads.

| BS 6446: 1997 | Specification for manufacture of glued structural components of timber and wood based panel products. |

BS 8103

BS 8103: Part 1: 1995	Code of practice for stability, site investigation, foundations and ground floor slabs for housing.
BS 8103: Part 2: 1996	Code of practice for masonry walls for housing.
BS 8103: Part 3: 1996	Code of practice for timber floors and roofs for housing.
BS 8103: Part 4: 1995	Code of practice for suspended concrete floors for housing.

BS 6566:	Plywood.
CP3: Chapter V: Part 2: 1972	Amd 4952, 5152, 5343, 6028 and 7908, Sept 1993. Wind loads.
DD ENV 1995-1-1: 1994	Amd 9148, July 1996. Eurocode 5: Design of timber structures.
DD 239: 1998	Recommendations for preservation of timber.
BS EN 338: 1998	Structural timber – Strength classes.
BS EN 380: 1993	Timber structures – Test methods – General principles for static load testing.
BS EN 385: 1995	Finger jointed structural timber.
BS EN 386: 1995	Glued laminated timber.
BS EN 518: 1995	Structural timber – Grading – Requirements for visual strength grading standards.
BS EN 519: 1995	Structural timber – Grading – Requirements for machine strength graded timber.
BS EN 1313-1: 1997	Round and sawn timber – Permitted deviations and preferred sizes. Part 1. Softwood sawn timber.
BS EN 20898 (BS EN 20898-1: 1992)	Mechanical properties of fasteners. Amd 7300, July 1992. Bolts, screws and studs.

Construction (Design and Management) Regulations 1994
Construction (Lifting Operations) Regulations 1961
Construction (Health, Safety and Welfare) Regulations 1996
Health and Safety at Work Act 1974
Management of Health and Safety at Work Regulations 1992
NHBC Standards Pitched Roofs – Chapter 7.2
Provision and Use of Work Equipment Regulations 1992
The Building Regulations 1991
The Manual Handling Operations Regulations 1992

Others

IS 193
Irish Statutory Instrument No. 138 of 1995   Safety, Health and Welfare at Work (Construction) Regulations 1995
The Building Regulations (Northern Ireland) 1994
The Building Standards (Scotland) Regulations 1990

TRA AGREED PROCEDURES

Useful References

TRADA Technology Ltd, Chiltern House, Stocking Lane, Hughenden Valley, High Wycombe, Bucks, HP14 4ND, UK
  Telephone: 01494 563091.   Fax: 01494 565487.
  Web site: http://www.tradatechnology.co.uk   E-mail: information@ttlchiltern.co.uk

Regional offices
TRADA Technology Ltd, Stirling Enterprise Park, John Player Building, Players Road, Stirling, FK7 7RS.
  Telephone: 01786 462122.   Fax: 01786 474412.
TRADA Technology Ltd, Templeborough Enterprise Park, Bowbridge Close, Rotherham, South Yorkshire, S60 1BT.
  Telephone: 01709 720215.   Fax: 01709 720178.

Publications
TRADA produce a wide range of technical publications covering all aspects of the use of both softwoods and hardwoods in the construction industry, including, of course, roof structures. The information is available both in CD-ROM form and as individual publications. The reader is recommended to obtain their publications list for further information.

Canadian Plywood Association, c/o Cofi, Suite 8, St Albans House, 40 Lynchford Road, Farnborough, Hants, GU14 6EF.
  Telephone: 01252 522545.   Fax: 01252 522546.
  Sundry publications and information on North American plywoods and timbers.

APA – The Engineered Wood Association, The Business Exchange, 39–41 Hinton Road, Bournemouth, Dorset, BH1 2EF.
Telephone: 01202 299277.   Fax: 01202 291505.   E-mail: apa-london@apawood.org
Sundry publications mainly concerning the use of American manufactured plywoods.

TRA – Trussed Rafter Association, Farndale, 31 Station Road, Sutton-cum-Lound, Retford, Notts, DN22 8PZ.
Telephone/Fax: 01777 869281.   E-mail: tra@pg/a.demon.co.uk
Trussed rafter technical manual.

TECHNICAL LITERATURE AND MANUALS

The Trussed Rafter Manual. Gang-Nail Systems Ltd, Christy Estate, Ivy Road, Aldershot, Hampshire, GU12 4XG.
Telephone: 01252 334691.   Fax: 01252 334562.   Web site: http://www.gangnail.co.uk

SpaceJoist leaflet. Gang-Nail Systems Ltd, details as above.

The World of Roof Technology. MiTek Industries Ltd, MiTek House, Grazebrook Industrial Park, Peartree Lane, Dudley, West Midlands, DY2 0XW.
Telephone: 01384 451400.   Fax: 01384 451417.   Web site: www.mitek.co.uk

Posi-Strut, Metal Web Truss System Engineering Manual. MiTek Industries, details as above.

The Trussed Rafter Guide. Alpine Ltd, Sandy Farm, Sands Road, The Sands, Nr Farnham, Surrey, GU10 1PX.
Telephone: 01252 781170.   Fax: 01252 782972.

Builders Fixing Guide. Alpine Ltd, details as above.

Technical Manual. Wolf Systems Ltd, Shilton Industrial Estate, Bulkington Road, Shilton, Coventry, CV7 9YJ.
Telephone: 01203 602303.   Fax: 01203 602243.
E-mail: .100621.2122@compuserve.com   Web site: www.wolfsystem.co.uk

Give Life to your Loft. The Velux Company Ltd, Woodside Way, Glenrothes East, Fife, KY7 4ND.
Telephone: 01592 772211.   Fax: 01592 771839.
E-mail: enquires@velux.co.uk   Web site: http://www.velux.com

Builders fixings literature. Simpson Strong-Tie, Winchester Road, Cardinal Point, Tamworth, Staffs, B78 3HG.
Telephone: 01827 255600.   Fax: 01827 255616.

BOOKS

Building in England – Sulzman.
   History of building development including roof structures.

Concise Encyclopaedia of Architecture – Martin and Briggs.
   Architectural styles.

Roofing Ready Reckoner – R. Goss.
   Practical cutting tables for timber roof components.

Roofs and Roofing – H.W. Harrison.
   Performance diagnosis, maintenance, repair and avoidance of defects.

Roofs and Roofing – Design and Specification Handbook – D.T. Coates.
   Roof coverings, insulation, acoustics and other technical properties.

Site Carpentry – C.K. Austen.
   Practical site carpentry.

The Building Regulations Explained and Illustrated – V. Powell-Smith and M.J. Billington.

Timber Designers' Manual – J.A. Baird and E.C. Ozelton.
   The structural design of timber components and structures.

Index